SpringerBriefs in Applied Sciences and Technology

PoliMI SpringerBriefs

W0079793

Series Editors

Barbara Pernici, DEIB, Politecnico di Milano, Milano, Italy

Stefano Della Torre, DABC, Politecnico di Milano, Milano, Italy

Bianca M. Colosimo, DMEC, Politecnico di Milano, Milano, Italy

Tiziano Faravelli, DCHEM, Politecnico di Milano, Milano, Italy

Roberto Paolucci, DICA, Politecnico di Milano, Milano, Italy

Silvia Piardi, Design, Politecnico di Milano, Milano, Italy

Gabriele Pasqui , DASTU, Politecnico di Milano, Milano, Italy

Springer, in cooperation with Politecnico di Milano, publishes the PoliMI Springer-Briefs, concise summaries of cutting-edge research and practical applications across a wide spectrum of fields. Featuring compact volumes of 50 to 125 (150 as a maximum) pages, the series covers a range of contents from professional to academic in the following research areas carried out at Politecnico:

- Aerospace Engineering
- Bioengineering
- Electrical Engineering
- Energy and Nuclear Science and Technology
- Environmental and Infrastructure Engineering
- Industrial Chemistry and Chemical Engineering
- Information Technology
- Management, Economics and Industrial Engineering
- Materials Engineering
- Mathematical Models and Methods in Engineering
- Mechanical Engineering
- Structural Seismic and Geotechnical Engineering
- Built Environment and Construction Engineering
- Physics
- Design and Technologies
- Urban Planning, Design, and Policy

Simone Garatti

Editor

Special Topics in Information Technology

POLITECNICO
MILANO 1863

Editor
Simone Garatti
Dipartimento di Elettronica,
Informazione e Bioingegneria
Politecnico di Milano
Milan, Italy

ISSN 2191-530X   ISSN 2191-5318   (electronic)
SpringerBriefs in Applied Sciences and Technology
ISSN 2282-2577   ISSN 2282-2585   (electronic)
PoliMI SpringerBriefs
ISBN 978-3-031-80267-6   ISBN 978-3-031-80268-3   (eBook)
https://doi.org/10.1007/978-3-031-80268-3

This work was supported by Politecnico di Milano.

This Springer imprint is published by the registered company Springer Nature Switzerland AG
The registered company address is: Gewerbestrasse 11, 6330 Cham, Switzerland

If disposing of this product, please recycle the paper.

Preface

It is with great pleasure and pride that I present this volume *Special Topics in Information Technology*, a collection of 12 outstanding contributions from Ph.D. students who have recently completed their doctoral studies in IT at the Dipartimento di Elettronica, Informazione e Bioingegneria of the Politecnico di Milano. These works have been carefully selected by the Faculty Board of the Ph.D. program in Information Technology from about 70 candidates and stand as a testament to the excellent research and innovation conducted in our department.

The Ph.D. program in Information Technology at the Politecnico di Milano has been in operation for about 40 years. It is built upon four main areas of study: computer science and engineering, electronics, systems and control, and telecommunications. This structure enables the program to cover an extensive range of research domains within the IT field. The aim is to promote research excellence by nurturing the development of innovative methodologies and cutting-edge technologies, and to prepare new generations of researchers and professionals, who will advance academic knowledge and drive innovation in industry and other sectors in the future. The authors of the contributions featured in this volume surely embody this spirit of excellence.

The present collection offers an exciting overview of the most significant research efforts undertaken by our Ph.D. students who graduated in the academic year 2023/2024. While grounded in rigorous scientific methods, the contributions are written in a way that makes the content accessible to a broad audience, and I am confident that both non-specialists and people with a more technical background will be able to appreciate the groundbreaking advancements achieved by our Ph.D. students. Each contribution reflects the diversity and depth of the four areas that define our Ph.D. program and showcases the high caliber of the research carried out at our institution. Whether theoretical or application-oriented, the presented works highlight both the fundamental and interdisciplinary nature of Information Technology and underscore its pivotal role in shaping modern society.

To conclude, I would like to express my gratitude to the contributing authors for their timely and enthusiastic participation in the creation of this volume. It is my hope that it will serve as both a record of their achievements and an inspiration for future generations of researchers.

Milan, Italy   Simone Garatti
October 2024

Contents

Computer Science and Engineering

Viral Data Integration and Knowledge Discovery Methods for Current and Future Pandemics

Tommaso Alfonsi ⓘ

Abstract Viral genomics is an interesting and challenging field of science. The vast amount of data, coupled with the intrinsic variability of viruses, demands robust data management and computational methods that support domain experts in studying this complex domain. This Chapter addresses the demand for data and knowledge integration as a way to analyse and discover new insights on SARS-CoV-2 and other viruses, even through artificial intelligence. Finally, a novel method for detecting recombination events in RNA viruses is presented. This method offers significant advantages over existing approaches and represents a valuable resource for public health preparedness. Overall, this work contributes significantly to viral genomics by addressing important challenges in data integration, knowledge modeling, and recombination detection.

1 Integration of Viral Data Sources

Following the outbreak of the SARS-CoV-2 pandemic, several organizations and research groups created online tools and resources to help study viral infections. Among the most relevant are the COVID-19 Analysis Research Database [1] (CARD) at the Children's Hospital of Los Angeles (CHLA), 2019nCoVR [2] and CoV-Seq [3]. They all rely on publicly available sequences collected from the most important database institutions: GISAID [4] (Global Initiative on Sharing All Influenza Data), NCBI [5] (National Center for Biotechnology Information), and CNGB [6] (China National GeneBank). Each of the integrated viral databases presented above is supported by a sequence integration pipeline that accomplishes duplicate sequence removal, feature annotation, variant calling, and (except for CARD) variant effect prediction. Despite some of them providing interesting additional features (See [7] for an in-depth comparison), they show some limitations in generality and data availability. Indeed, these resources do not collect samples of viral species other than SARS-CoV-2 nor other relevant region annotations. An example of these is epitope regions, whose

T. Alfonsi (✉)
Dipartimento di Elettronica, Informazione e Bioingegneria, Politecnico di Milano, Milano, Italy
e-mail: tommaso.alfonsi@polimi.it

© The Author(s) 2025
S. Garatti (ed.), *Special Topics in Information Technology*,
PoliMI SpringerBriefs, https://doi.org/10.1007/978-3-031-80268-3_1

stability is of vital importance for vaccine effectiveness and pathogen recognition by the host. Moreover, the possibility of expressing and combining filters on sequences' metadata and the viral genome (both at the nucleotide or amino acid level) is limited or absent. Further comparisons are made difficult by the low amount of details about the internal database structure chosen for representing the isolate's data. To address these limitations, we developed ViruSurf [8], a highly scalable, extensible data repository for viral sequences. We extended it with epitope data that can be queried and analyzed in a dedicated interface called EpiSurf [9]. Finally, we built the mutation and epitope data analysis tool VirusViz [10], which is interconnected with the previous resources. To support the work of integration, annotation of sequences and mutation extraction, an Extract-Transform-Load (ETL) pipeline [7] has been designed to integrate data from several sources of viral sequences periodically. The pipeline is composed of layers for downloading data from the sources, extracting information and homogenizing its content, and finally, loading it into a central repository. The first layer includes source-specific modules responsible for collecting the data from the most important viral sequence deposition resources: NCBI, the National Microbiology Data Center (NMDC), COG-UK [11] and GISAID. The original data is mapped to a relational model (based on the Virus Conceptual Model [12]), allowing a representation of the virus under different perspectives (biological, technological, analytical and organizational) and its seamless extension with other concepts. A transformation layer ensures that similar entities are reconciled so queries can seamlessly operate on data originating from different sources. The pipeline is completed by a query optimization layer that computes accessory data structures to increase query performance on such a huge viral sequence repository. Subsequently, a dedicated set of pipeline modules has been developed to add epitope annotations from the Immune Epitope Database (IEDB [13]) and support their integration with viral sequences.

Following a substantial increase in SARS-CoV-2 sequence submissions to NCBI GenBank (47,193 in 2020 to 3,018,196 by 2021), the ViruSurf platform underwent optimizations to ensure efficient data integration and performance. Previously, ViruSurf relied on the creation of periodic, updated databases. This approach was replaced with a single, continuously updated repository containing snapshots. This optimization improved efficiency and reduced idle times. Additionally, dedicated server infrastructure was implemented to handle the computational demands of the ViruSurf pipeline, minimizing resource consumption on the interface machines (ViruSurf, EpiSurf, VirusViz). Furthermore, automating snapshot management tasks (i.e., image creation, transfer, load, and finalization) significantly streamlined the process. These optimizations effectively addressed scalability and performance concerns associated with integrating large-scale viral sequence data.

The ViruSurf pipeline currently generates an integrated sequence repository comprised of two databases. The first database contains over 5.4 million SARS-CoV-2 sequences obtained from GISAID. The second database houses sequences from NMDC, COG-UK, GenBank, and RefSeq, encompassing 3.7 million SARS-CoV-2 sequences and 35,000 sequences from other viral species.

The data generated by such a pipeline was fundamental for building the web applications ViruSurf, EpiSurf and VirusViz and represents a significant contribution to making viral data more accessible and insightful for the global scientific community.

2 Integration of Viral Knowledge

The ongoing deposition of SARS-CoV-2 sequences into public databases facilitates continuous monitoring of viral evolution by international organizations. Several research studies link mutational processes to the emergence of more transmissible variants, immune escape, and altered disease course. Understanding these dynamics is crucial for understanding the SARS-CoV-2 pandemic. However, this effort is hampered by several factors. One issue is the use of inconsistent terminology, as the lack of standardized terms and concepts across research groups hinders the efficient sharing of data and knowledge. Another important aspect is related to heterogeneity and temporal evolution, as viral sequences and associated information exhibit inherent variability and change over time, making knowledge management challenging. Finally, existing resources often lack the ability to connect disparate knowledge domains related to SARS-CoV-2.

Addressing these challenges, we introduce CoV2K [14], a unifying knowledge model about SARS-CoV-2. CoV2K offers an abstract representation of SARS-CoV-2 information using entities and relationships, promoting consistent terminology and facilitating the integration of diverse data sources. The entities described by the CoV2K model are variants clinical effects of variants and amino acid changes, scientific literature providing evidence for the effects, nucleotide and amino acid changes, gene and protein annotations and finally, chemical properties of individual residues. Entities are also linked to massive datasets of sequence data (ViruSurf [8]) and epitope data (EpiSurf [9]) as discussed in Sect. 1.

We filled the content of entities and relationships using multiple sources. More specifically, CoVariants.org [15] and PHE [16] are used for variant names and their characteristic mutations. COG-UK Mutation Explorer [11] provides a comprehensive list of the effects of SARS-CoV-2 mutations. Effects of variants are extracted from ECDC reports [17] and the CoVariants web page [15]. Entities corresponding to nucleotide and amino acid change include the mutations found in any variant characterizations and sequence of ViruSruf. NCBI's gene and protein annotations for the reference sequence NC_045512 [18] are represented in the corresponding entities of the model. Protein domains are downloaded from UniProtKB [19]. Properties of residues have been extracted from the Amino Acid Explorer of NCBI Structures [20] in Sept. 2020 and checked against another authoritative reference [21] as the NCBI resource has been discontinued.

In order to extract information from the above sources, established Extract, Transform, Load (ETL) methodologies have been used. In addition, a final harmonization stage concludes the integration process. During the first stage (extraction), data is retrieved from the designated sources using source-specific parsers to identify relevant information within the source data structures. In the transformation stage, the extracted data is mapped to a unified MongoDB representation. Entities are converted to "collections" and data instances become "documents", adhering to the standard terminology of NoSQL document databases. The transformed collections of documents are directly inserted into a MongoDB instance (loading stage) without duplicate checking, as this operation is postponed to the next stage. The final stage (harmonization) employs MongoDB's aggregation framework to normalize collections. This process integrates and eliminates duplicate documents, standardizes attribute vocabularies, and computes missing or implicit information (e.g., inferring source organization of a variant based on naming pattern). This setting reduces code maintenance issues, as the harmonization procedure is fully implemented through a declarative approach.

The unified representation of knowledge is provided to the broader public by the CoV2K-API [14], a RESTful API implementing one endpoint for each model's entity named as the entity itself. When called without optional parameters, the endpoints return all the objects of the associated entity. Alternatively, endpoints accept as input parameters either the ID of a single object (if known from a previous query) or a list of query parameters. In the former case, the API returns the full description of the referenced object; in the latter, a list of matching object IDs is returned. User queries that involve chaining multiple entities can be conveniently written through the special endpoint */combine*, for example, as in http://gmql.eu/cov2k/api/combine/evidences/effects?aa_positional_change_id=S:L452R, which extracts the scientific literature (evidences) reporting effects on the Spike mutation L452R.

In conclusion, CoV2K offers a robust query system that facilitates user exploration of interconnected data and knowledge domains related to SARS-CoV-2. The knowledge sources within CoV2K are meticulously chosen for their accuracy and adherence to rigorous update schedules, ensuring users access the most current information. Furthermore, the implemented ETL and harmonization pipelines are designed for seamless integration of additional data sources in future. On the data side, CoV2K leverages two substantial databases: ViruSurf and EpiSurf. As of early 2022, ViruSurf houses approximately 4M sequences from GenBank and COG-UK, enriched with nucleotide mutations and corresponding amino acid changes. EpiSurf, in turn, provides a curated collection of roughly 22K epitope annotations on SARS-CoV-2.

We believe that CoV2K represents a valuable resource in support of viral investigators by cleverly integrating diverse knowledge domains connected to the SARS-CoV-2, fostering the discovery of novel insights. The graph-shaped model also facilitates the use of artificial intelligence methods on CoV2K, for example, by means of knowledge graph analysis tools. One notable example is also provided in [22], where we employed a logical declarative language to model intensional knowledge (i.e., the rules or entity relationships) and express the extensional knowledge (i.e., facts). Then, the reasoner applies the rules to the known facts and derives novel extensional knowledge. Four use cases of such an approach are presented in [22].

3 Recombinant Detection

Since the beginning of the SARS-CoV-2 pandemic, more than 130K distinct nucleotide mutations have accumulated globally. Their combination defines the genomic and functional characteristics of a viral sequence. As all such sequences ultimately descend from a single ancestor, it is possible to place them in a diverging tree-like data structure, called a phylogenetic tree, that shows the evolution of the virus from a genomic and epidemiological standpoint. Phylogenists name the tree's nodes as variants—or lineages or clades—i.e., groups of sequences sharing similar genomic features. So, it's possible to tell the similarity and origin of a new variant starting from the pre-existing ones. Despite the primary role of phylogenetic analysis, such an approach relies on the evolutionary model of a branching tree, but the suitability of this model to describe the history of SARS-CoV-2 was questioned when recombination was first observed in this virus [23].

Recombination is the exchange of genetic material between two or more viral strains co-circulating in the same host. Through recombination, a virus acquires a contiguous portion of mutations of another sequence abruptly, leading to greater diversification and faster evolution.

Despite, in the first three years of the SARS-CoV-2 pandemic, only low levels of recombination were noted, this phenomenon is extremely current, and several recombination events have been recognized as the pandemic continued spreading [24, 25]. Recombination is a critical evolutionary mechanism in SARS-CoV-2 and other viruses, and it may have severe clinical and epidemiological implications as it is thought to increase the probability of cross-specie transmission, enhanced virulence, host immune evasion, and the development of resistance to antivirals [26]. Recognition of recombinant sequences is, therefore, vital for the fight against SARS-CoV-2 and future pandemics.

As recombination disrupts the branching tree model, it impairs the application of phylogenetic approaches for tracing contagion, viral evolution, and ultimately genomic surveillance of pathogens [27]. Therefore, an effective method to detect recombinants on a pandemic-scale dataset is paramount for guiding the scientific community in preventing and mitigating future pandemic threats. Targeting this ambitious goal, we have developed a novel data-driven approach for detecting recombinant sequences, which is not reliant on phylogeny, is highly precise, and applicable to vast viral datasets, including SARS-CoV-2, Mpox, Dengue, West Nile virus and others.

Our method, RecombinHunt [28], classifies a series of mutations within a genomic sequence and detects the genomic coordinates where a discontinuity in the classification arises—a genomic position known as "breakpoint". Here, classification is a general term referring to the assignment to a known variant depending on the variant nomenclature system in use. The assignment of a portion of the genome to a specific variant depends on the likelihood of observing a specific series of mutations. RecombinHunt estimates the likelihood of single mutations knowing its global frequency (i.e., in all viral sequences) and relative frequencies $f_1, f_2, ..., f_V$ for V variants. For any candidate variant and coordinate range on the target sequence, RecombinHunt computes the cumulative likelihood considering the series of mutations included in the range and detects where such likelihood starts decreasing. Indeed, the maximum cumulative likelihood for a non-recombinant sequence is found at the widest possible coordinates range on the target sequence and for the most likely variant assignment. For a recombinant sequence, instead, the maximum score occurs at the breakpoint position. The breakpoint is the genomic coordinate that divides two portions of a recombinant genome inherited by different parents. Once the breakpoint position has been identified, the first parental variant is found as the one with the highest score, and the cumulative likelihood is recomputed on the remaining portion of the genome to identify the other parental candidate variant.

RecombinHunt detects single breakpoint recombinant genomes (i.e., sequences made by only two parental genome portions) and double breakpoint recombinant genomes—i.e., where a portion of one parental sequence is placed between two portions of the other parental genome. These recombination patterns are also respectively called $L1+L2$ and $L1+L2+L1^{opp}$ for brevity, where $L1$ and $L2$ refer to two generic parental variants and ^{opp} refers to the opposite end of $L1$. These are the two most common recombination patterns due to the functional limitations on the genome assembly,[1] and because the simultaneous infection by three or more viruses is less likely. The workflow of RecombinHunt is such that the non-recombinant and recombinant options are nevertheless computed. Then, the model with the lowest entropy according to the Akaike Information Criterion [29] is chosen as the most likely explanation. This choice reflects the observation that positive selection is most often the right answer, unlike recombination. Recombination is a rare phenomenon, and $L1+L2+L1^{opp}$ recombination patterns are even more rare than $L1+L2$. So, if a non-recombinant model explains sufficiently well the mutations observed in a sequence, it will be the preferred answer of RecombinHunt.

The method has been thoroughly evaluated in [28] and achieved outstanding results both on SARS-CoV-2 and Mpox viruses. As demonstrated in [28], RecombinHunt scored a five-fold increase in sensitivity compared to the state-of-the-art on real recombinant cases.

To evaluate the sensitivity and specificity of RecombinHunt more reliably, we also simulated 10,500 sequences equally distributed among the classes: non-recombinant, single breakpoint recombinant and double breakpoint recombinant sequences. Then, we added random non-characteristic mutations (i.e., noise) at increasing quantities from 0 to 30.[2] In these tests, we achieved almost perfect results. Considering sequences with up to 10 added noise mutations, we measured more than 95.6% sensitivity and less than 1.2% false positives rate.

---

[1] The structure of a recombinant genome is not different from that of a non-recombinant genome. They must encode for the same proteins, have similar lengths, and have the same arrangement of genome parts to preserve the viral functions. Therefore, parental genomes cannot be randomly assembled.

[2] The generated recombinant sequences had 60 mutations on average compared to the reference. So, adding 30 mutations as noise corresponds to adding ~50% of noise in the input.

RecombinHunt has been designed to work in a big-data context, a scenario where the applicability of many other methods is limited. On small datasets, the main limit of RecombinHunt is the potentially inaccurate estimate of variant characterizations, which, in turn, depends on the accuracy of mutation frequencies and the chosen characterization threshold.[3] In [28], we illustrated two approaches to determine the minimum dataset size for obtaining a stable characterization. In both cases, we conclude that the answer depends on the quality of the data and the sequence classification system used. Having a sound variant classification system with few noise mutations added per sequence (e.g., below 15), 5 sequences per lineage are enough to derive a correct characterization (for example, of PANGO lineage AY.44). But the variant characterization of PANGO lineages AY.45 and Mpox lineage B.1 required more than a hundred sequences before achieving a stable characterization. This example outlines the importance of the input data quality on the applicability of RecombinHunt.

Taking advantage of the generality of RecombinHunt, we tested the method on the Mpox virus. After building the variant characterizations and frequency data for the 21 Mpox lineages, we identified more than four hundred recombinant sequences. Most importantly, we confirmed 5 out of 8 double breakpoint recombinant sequences that were previously identified by an ad-hoc study on Mpox [30].

Finally, we demonstrated in [28] the possibility of using RecombinHunt as a prediction method for detecting novel recombinants. In particular, the method was tested on 24 sequences initially designated as non-recombinant in GISAID. Out of these, we identified 19 recombinants with parental lineages and breakpoint positions compatible with the recombinant designations XCA and XCB, which were introduced only subsequently. Later, the same sequences were also labelled recombinant in GISAD and assigned to XCA or XCB lineages.

Our findings collectively demonstrate the high accuracy and reliability of RecombinHunt, marking a significant advancement in the detection of recombinant viruses. This method exhibits broad applicability to existing nucleotide mutation collections for various viral species, including Zika and Ebola. Consequently, RecombinHunt facilitates the timely detection of recombinant viral genomes in ongoing and future outbreaks. As a future direction, we aim to implement the method for influenza virus detection with the necessary adaptations for multi-segmented genomes.

4 Conclusions

The emergence of SARS-CoV-2 highlighted the critical role of genomic surveillance, the study of pathogen evolution through genome sequencing. The heterogeneity of the viral data presents additional complexity for scientists wanting to focus on the biological intricacies of viruses. This Chapter illustrates a general and extensible

---

[3] For SARS-CoV-2, a 75% characterization threshold was used. The higher the threshold, the higher the confidence level needed for a mutation to be part of the characteristic set of mutations.

approach to viral data and knowledge integration and presents a novel method for recombinant detection that significantly improves the state-of-the-art. In conclusion, this thesis addresses a number of open challenges in viral genomics by proposing novel methods that can effectively contribute to genomic surveillance practices and improve global health preparedness against future pandemics.

References

1. Shen, L., Maglinte, D., Ostrow, D., Pandey, U., Bootwalla, M., Ryutov, A., Govindarajan, A., Ruble, D., Han, J., Triche, T.J., Bard, J.D., Biegel, J.A., Judkins, A.R., Gai, X.: Children's Hospital Los Angeles COVID-19 Analysis Research Database (CARD)-A Resource for Rapid SARS-CoV-2 Genome Identification Using Interactive Online Phylogenetic Tools. bioRxiv (2020). https://doi.org/10.1101/2020.05.11.089763
2. Lu, G., Moriyama, E.N.: 2019ncovr-a comprehensive genomic resource for SARS-COV-2 variant surveillance. Innov. **2**(4), 100150 (2021)
3. Liu, B., Liu, K., Zhang, H., Zhang, L., Bian, Y., Huang, L.: CoV-Seq: SARS-CoV-2 Genome Analysis and Visualization. bioRxiv (2020). https://doi.org/10.1101/2020.05.01.071050
4. Shu, Y., McCauley, J.: GISAID: global initiative on sharing all influenza data–from vision to reality. Eurosurveillance **22**(13) (2017)
5. NCBI Resource Coordinators: Database resources of the National Center for Biotechnology Information. Nucl. Acids Res. **46**(D1), D8–D13 (2017)
6. Chen, F.Z., You, L.J., Yang, F., Wang, L.N., Guo, X.Q., Gao, F., Hua, C., Tan, C., Fang, L., Shan, R.Q., et al.: Cngbdb: china national Genebank database. Yi Chuan= Hereditas **42**(8), 799–809 (2020)
7. Alfonsi, T., Pinoli, P., Canakoglu, A.: High performance integration pipeline for viral and epitope sequences. BioTech **11**(1), 7 (2022)
8. Canakoglu, A., Pinoli, P., Bernasconi, A., Alfonsi, T., Melidis, D.P., Ceri, S.: ViruSurf: an integrated database to investigate viral sequences. Nucl. Acids Res. **49**(D1), D817–D824 (2021)
9. Bernasconi, A., Cilibrasi, L., Al Khalaf, R., Alfonsi, T., Ceri, S., Pinoli, P., Canakoglu, A.: EpiSurf: metadata-driven search server for analyzing amino acid changes within epitopes of SARS-CoV-2 and other viral species. Database **2021** (2021)
10. Bernasconi, A., Gulino, A., Alfonsi, T., Canakoglu, A., Pinoli, P., Sandionigi, A., Ceri, S.: VirusViz: comparative analysis and effective visualization of viral nucleotide and amino acid variants. Nucl. Acids Res. **49**(15), e90 (2021)
11. The COVID-19 Genomics UK (COG-UK) consortium. An integrated national scale SARS-CoV-2 genomic surveillance network. Lancet Microbe **1**(3), e99 (2020)
12. Bernasconi, A., Canakoglu, A., Pinoli, P., Ceri, S.: Empowering virus sequence research through conceptual modeling. In: Dobbie, G., Frank, U., Kappel, G., Liddle, S.W., Mayr, H.C. (eds.) Conceptual Modeling, pp. 388–402. Springer International Publishing, Cham (2020)
13. Vita, R., Mahajan, S., Overton, J.A., Dhanda, S.K., Martini, S., Cantrell, J.R., Wheeler, D.K., Sette, A., Peters, B.: The immune epitope database (IEDB): 2018 update. Nucl. Acids Res. **47**(D1), D339–D343 (2019)
14. Alfonsi, T., Al Khalaf, R., Ceri, S., Bernasconi, A.: CoV2K model, a comprehensive representation of SARS-CoV-2 knowledge and data interplay. Sci. Data **9**(1), 1–12 (2022)
15. Hodcroft, E.B.: CoVariants: SARS-CoV-2 Mutations and Variants of Interest (2022). Accessed 20 Jul. 2024

16. Public Health England: COVID-19 variants: genomically confirmed case numbers (2022). Accessed 20 Jul. 2024
17. European Centre for Disease Prevention and Control: SARS-CoV-2 variants of concern (2022). Accessed 20 Jul 2024
18. Fan, W., Zhao, S., Bin, Yu., Chen, Y.-M., Wang, W., Song, Z.-G., Yi, H., Tao, Z.-W., Tian, J.-H., Pei, Y.-Y., et al.: A new coronavirus associated with human respiratory disease in china. Nature **579**(7798), 265–269 (2020)
19. The UniProt Consortium: Uniprot: the universal protein knowledgebase in 2021. Nucl. Acids Res. **49**(D1), D480–D489 (2021)
20. NCBI: NCBI Structures Amino Acid Explorer Resource. Accessed 23 Sept. 2020
21. Barrett, G.: Chemistry and Biochemistry of the Amino Acids. Springer Science & Business Media (2012)
22. Alfonsi, T., Luigi, B., Bernasconi, A., Ceri, S., et al.: Expressing biological problems with logical reasoning languages. In: Ceur Workshop Proceedings, vol. 3229, pp. 1–15. CEUR-WS (2022)
23. Van Insberghe, D., Neish, A.S., Lowen, A.C., Koelle, K.: Recombinant SARS-COV-2 genomes circulated at low levels over the first year of the pandemic. Virus Evol. **7**(2), veab059 (2021)
24. Mohapatra, R.K., Kandi, V., Tuli, H.S., Chakraborty, C., Dhama, K.: The recombinant variants of SARS-COV-2: concerns continues amid COVID-19 pandemic. J. Med. Virol. **94**(8), 3506 (2022)
25. World Health Organization: Tracking SARS-CoV-2 variants. Accessed 20 Jul. 2024
26. Focosi, D., Maggi, F., Franchini, M., McConnell, S., Casadevall, A.: Analysis of immune escape variants from antibody-based therapeutics against COVID-19: a systematic review. Int. J. Mol. Sci. **23**(1), 29 (2021)
27. Schierup, M.H., Hein, J.: Consequences of recombination on traditional phylogenetic analysis. Genetics **156**(2), 879–891 (2000)
28. Alfonsi, T., Bernasconi, A., Chiara, M., Ceri, S.: Data-driven recombination detection in viral genomes. Nat. Commun. **15**(1), 3313 (2024)
29. Akaike, H.: A new look at the statistical model identification. IEEE Trans. Autom. Control **19**(6), 716–723 (1974)
30. Yeh, T.-Y., Hsieh, Z.-Y., Feehley, M.C., Feehley, P.J., Contreras, G.P., Su, Y.-C., Hsieh, S.-L., Lewis, D.A.: Recombination shapes the 2022 monkeypox (mpox) outbreak. Med **3**(12), 824–826 (2022)

Learning Optimal Equilibria and Mechanisms Under Information Asymmetry

Federico Cacciamani (ID)

1 Introduction

Over the last few decades, scientific research in the field of *artificial intelligence* (AI) has been capable of achieving outstanding results in modeling optimal agents' behavior in many situations of strategic interaction. Such results include, but are not limited to, the achievement of *human* and *superhuman* performances in increasingly difficult recreational games like *Chess* [6], *Go* [16], *No-limit Texas Hold'em Poker* [2, 13] and *Starcraft II* [17]. In all those cases, the development of game-theoretic reasoning served as a fundamental building block for obtaining the final result. Indeed, differently than standard settings that deal with learning to behave in a *fixed* environment, complex scenarios like the ones listed above require an additional bit of reasoning, which accounts for the fact that there might be other agents that can *strategically* adapt their behavior to the one of the others. This aspect caused an increasing interest revolved around the subject of *algorithmic game theory*, which is indeed concerned with the computational study of the interaction between different strategic agents in complex real-world scenarios. At the core of this subject, there are the concepts of *game*, which is the mathematical model of the strategic interaction, and *equilibrium*, which represents a solution of the game. The foundation of the field can be traced back to the seminal work by Nash [15], that introduced the concept of *Nash Equilibrium* (NE), to which other notions of equilibria followed (see e.g., the notion of *Correlated Equilibrium* introduced by Aumann [1]). Since then, researchers in the field of algorithmic game theory have been concerned with finding efficient algorithms to compute or approximate different notions of equilibrium, allowing the deployment of such strategies in many real-world scenarios like the aforementioned ones.

F. Cacciamani (✉)
Politecnico di Milano, Piazza Leonardo da Vinci 32, Milan, Italy
e-mail: federico.cacciamani@polimi.it

© The Author(s) 2025
S. Garatti (ed.), *Special Topics in Information Technology*,
PoliMI SpringerBriefs, https://doi.org/10.1007/978-3-031-80268-3_2

Most of the research in the field has been focused around the development of scalable algorithms for finding NE in simple settings like two-players games in which the players have opposite goals (i.e., two-players zero-sum games). These games present nice mathematical properties that allowed the definition of a large number of different approaches to tackle the problem. Over the last years, the most popular approach has been undoubtedly the adoption of *learning algorithms*, i.e., algorithms that are capable of learning optimal strategies by means of samples of repeated interactions between the agents [11, 22]. The advantages of learning algorithms over standard optimization-based approaches are several, as they allow to frame the problem of computing equilibria in a decentralized way, and they are suitable to be combined with black-box models of the environment without the need for a prior knowledge of the game. These aspects are crucial when dealing with large games (as many real-world applications would require) that are too big to be tackled with optimization-based algorithms.

Despite the successes mentioned above, the expressive power of two-players zero-sum games is not enough to model most of the actual scenarios of strategic interaction that can be found in modern society. This is because the vast majority of those situations are inherently characterized by a large number of agents. Consider, for instance, the case of a ride-sharing platform that wants to improve its efficiency or the case of an investor who is interested in designing the best possible trading strategies for increasing her wealth. In both cases, there is a huge number of agents involved in the strategic interactions, and more powerful models than the one provided by two-players zero-sum games are needed. Unluckily, when the number of players in the game grows, most of the convenient properties characterizing two-players games are lost, and the problem of understanding which is the most relevant solution concept, as well as the problem of computing it, becomes significantly more complex. From a computational perspective, the problem gets even more complicated when we consider games in which the players hold private information which cannot be observed by the others.

In this manuscript, we aim to advance the state of the art in the directions sketched above, by studying the development of learning algorithms for different types of multi-agent systems. In particular, we study scenarios of sequential strategic interaction between multiple players, in which each of them holds some partial information on the state of the environment and, depending on the structure of the interaction, we investigate how this asymmetry of information influences the solution concept and how it can be leveraged in order to design *learning algorithms* capable of learning the optimal strategies. We differentiate between two cases that model whether the agents holding the information are *active*, i.e., they actively take actions in the game, or *passive*, i.e., they do not take actions in the game but can influence the behavior of other agents by strategically disclosing information.

2 Active Agents: Ex-ante Coordination

The first scenario that we investigate in this manuscript is the one in which there are many players that sequentially interact between them in a game in which each of them holds some partial information on the state of the environment. We study cases in which the nature of interaction is *mixed cooperative-competitive*. Thus, in some situations, the players might be interested in cooperating between them in order to enforce more favorable outcomes for them, while in other ones, their interests might collide. Depending on their objectives, the players might be interested in communicating some information to the other agents, either to better cooperate or to induce them to act conveniently. In this work, we mainly focus on cases in which the type of communication allowed is *ex-ante* (or *preplay*) *communication*, i.e., cases in which the players can only communicate before the beginning of the game and agree on some shared strategy, but cannot explicitly communicate during the execution of the game. Restricting the possible communication to the ex-ante one allows us to capture many real-world scenarios in which communication between agents is impossible during the interaction (for instance, due to environmental obstacles). Moreover, this kind of communication opens up a very interesting interpretation of the strategies of the players. In particular, the strategies can be thought of as communication protocols that, based on the *public information* available, associate specific meanings to actions. The concept of public information will indeed prove to be a fundamental concept in our study.

Formally, the preplay communication protocol is implemented by means of a *shared mediator*, which privately and sequentially issues communications, taking the form of *action recommendations*, to each player. Our objective, then, becomes to find optimal mediator strategies, which correspond to *optimal equilibria*, i.e., recommendation policies maximizing some function (e.g., the *social welfare*) such that all the players are incentivized to follow the recommendations issued by the mediator. More precisely, we tackle the problem of finding optimal equilibria from a *learning perspective*, i.e., we are interested in designing algorithms that, leveraging samples of sequential interactions between the agents, are capable of learning such optimal equilibria.

Depending on the types of players involved in the game, we consider different types of equilibria. In particular, we differentiate between cases in which players are divided in two *teams* with opposite objective and more general cases in which there is not any particular structure among the players. In the former case, the objective of our study will be to learn the so-called *team correlated equilibria* [8], while in the latter case we will be interested in learning optimal *Extensive-Form Correlated Equilibria* (EFCE) [18] and *Extensive-Form Coarse Correlated Equilibria* (EFCCE) [9], which have been established as the most relevant solution concepts for general-sum sequential games.

2.1   Team Games

The first problem that we tackle corresponds to the development of learning algorithms for equilibria in team games. The type of team games that we consider are
characterized by two teams of players that face each other in a zero-sum interaction
(i.e., they have colliding interests). An example of one such game is the game of
Bridge. As already mentioned, our model does not allow communication between
the agents during the interaction, but only admits *ex-ante* communication. In this way,
the players in the same team can agree on a joint strategy before the beginning of
the game and can subsequently convey information only by means of their behavior
in the game. In such a setting, previous research have established the *Team Maxmin
Equilibrium with correlation* (TMEcor) [5, 8, 10, 21] as the most appropriate solution concept. At an high level, the TMEcor can be formulated as the solution of a
saddle-point optimization problem, where the optimization is done by considering
the joint strategy sets of the two teams. Unfortunately, finding such an equilibrium
was shown to be NP-hard [8]. Nonetheless, despite the computational complexity
results, we believe that it is of uttermost importance to develop efficient algorithms
that allow to address this problem as this would enable a wide range of applications
for the theory. Furthermore, existent algorithms for computing a TMEcor are heavily based on the solution of complex optimization problems such as integer linear
programs and mixed integer linear programs [8, 10, 21], which, in practice, makes
the adoption of such techniques impossible in many real-world scenarios. Similar
to what happened for two-players zero-sum games, we believe that introducing a
learning-based framework for the TMEcor computation, which can potentially be
combined with highly scalable deep-learning techniques, could significantly increase
the potential related to the use of this game-theoretic concepts in real-world scenarios.
    Starting from this consideration, we develop a procedure that allows us to con-
vert the team game into a strategically equivalent two-players zero-sum game. Our
algorithm is based on the concept of *public information*, which captures the amount
of information that is public among members of the same team. More in detail, we
create a two-players zero-sum game, which we call *Team Public Information* (TPI)
game, in which the players in the same team are replaced by a meta-player called
coordinator. When the game reaches a certain decision node of a team player, the
corresponding coordinator plays. Instead of playing simple actions, the coordinator chooses *prescriptions*, which specify an action for each possible decision node
that is coherent with the public information available to the team players at that
specific point in the game. Then, after the choice of a prescription, the team player
chooses the action that corresponds to its actual private state, and the game continues
accordingly. Using this kind of coordination based on public information allows us
to mitigate the aspect that causes the high complexity of computing a TMEcor. In
particular, such complexity is due to the imperfect recallness of the team if seen as
a unique player. By considering only public information, we ensure that the coordinator maintains perfect recall, enabling the use of well-known learning algorithms
(see e.g., [22]) for finding a NE of the TPI game and the corresponding TMEcor in

the original game. Unfortunately, the TPI game has a size that grows exponentially with the size of the original game (but this is inevitable unless P=NP). To mitigate this issue, we take inspiration from the literature on two-players zero-sum games and introduce in the context of team games two techniques that proved fundamental to achieve superhuman performances in huge and very complex games like no-limit texas hold'em poker: *stochastic regret minimization* [12] and *subgame solving* [14]. In particular, stochastic regret minimization allows to reduce the computational burden of the single iteration of the learning algorithm via sampling techniques, while subgame solving algorithms are used to improve the performances of a given strategy only on a small portion of the game tree, without the need of traversing (or even knowing !!) the rest of the game. To the best of our knowledge, we introduce the first team-based versions of stochastic regret minimization and subgame solving, leveraging the strategy formulation proper of our TPI transformation.

The results mentioned above were published as [7, 19].

2.2 General-Sum Games

Then, we move to the study of how to compute optimal correlated equilibria in general-sum games. Correlated equilibria are formally modeled by means of a mediator that samples action recommendations for the players from a given probability distribution over the set of joint strategies, and sequentially issues such recommendations to them. A mediator strategy is an equilibrium if no player has incentive to deviate from what the mediator recommends. In such settings, due to the known connection between no-regret learning and convergence to correlated equilibria, the state-of-the-art approach consists of running decentralized no-regret learning dynamics for each player in the game, which guarantees convergence to the set of correlated equilibria of the game [11]. Such solution concepts are particularly significant to be used in complex systems e.g., *ride-sharing platforms* in which there is the need to incentivize specific behavior of the agents involved so to guarantee proper stability of the system. In many of such applications, however, it may be required to guarantee some sort of *fairness* for the solution reached, for instance because of legal requirements. In this sense, the use of decentralized learning dynamics in the way described above is no longer a viable option, since that approach does not offer any guarantee on the quality of the equilibrium reached.

In light of this undesired behavior, we address the problem of developing algorithms for learning *optimal* (i.e., maximizing some given objective function) EFCE and EFCCE. Also in this case, previous research showed that computing such equilibria constitutes an NP-hard problem [18]. For this reason, we are interested in casting this complex problem in a suitable learning framework that, similar to what we discussed for team games, would allow to bypass the complexity results by means of the generalization capabilities offered by deep learning algorithms. To the best of our knowledge, this is the first work addressing the problem of computing such optimal equilibria from a learning perspective.

To deal with these challenges, we frame the problem from the perspective of a mediator issuing action recommendations to the players. By adopting this centralized view, we were able to formulate a *primal-dual* based learning algorithm, which converges to an optimal correlated equilibrium. To this extent, we leverage a *sandboxed* type of interaction between the mediator and the agents, i.e., the mediator can interact with each agent separately, without the interference of the others. Then, starting from the lagrangian relaxation of the optimization problem for finding the optimal correlated equilibrium, we can formulate an interleaved learning dynamic between a *primal* regret minimizer, a *dual* regret minimizer and one regret minimizer for each player, guaranteeing that the average primal strategy approaches an optimal equilibrium as the time horizon grows.

The above results were published as [20].

3   Passive Agents: Information Acquisition

The scenario in which the agents holding information can not directly play actions to influence the outcome of the game but can only influence the behavior of other agents by strategically disclosing information models a problem called *information acquisition*. Due to the increasing decentralization of information that characterizes modern society, information acquisition represents a ubiquitous case of strategic interaction in the real world. We introduce a typical information acquisition problem with a simple example. Consider a portfolio manager who wants to learn the potential of a company to make an informed investment. The manager could *hire* multiple analysts to conduct separate researches on the same company, where each analyst spends effort to acquire information and produce a report. Then, the analysts disclose their reports to the manager who, based on the information received, decides whether or not to make the investment. In the general case, the analysts might have some private interests in the company and might want to forge their report in order to induce some specific action by the manager and have personal gain from that. The manager, instead, has interest in avoiding to act based on forged information and can pay the analysts to avoid such undesirable outcomes.

Formally, the information acquisition problem can be framed as a particular instance of the *principal-agent problem* in which the portfolio manager acts as the principal and the analysts act as the agents. In this model, the environment has some stochastic state which is hidden both to the principal and to the agents. The agents can choose among different effort levels, incurring in different costs, and receive some stochastic information on the state of the environment based on the effort level chosen. Then, they can report (possibly non-truthfully) their information to the principal, who aggregates the information received to select which action to take and can pay the agents based on the quality of their reports. We frame such a game-theoretical model into a typical online learning scenario in which the principal sequentially interacts with a set of agents in an unknown environment. The objective of our study revolves around answering the following question:

How can the principal learn to optimally coordinate the efforts of the agents and combine their reports in order to take the best possible action?

We answer the above question by differentiating between two distinct cases that amount to different capabilities of the principal. In particular, the ways in which the principal can influence the agents' behavior are essentially two: (i) by choosing a suitable action selection policy, and (ii) by choosing suitable payment schemes. We refer to the set of principal's policies as *mechanisms*. In this work, we study the two cases separately, focusing first on the case in which the principal does not have budget to issue payments to the agent and then to the case in which the action taken by the principal does not influence agents' utilities. The two particular cases present specific challenges that need to be faced to achieve our goal.

In order to leverage the feedback received from the interaction and learn the optimal mechanism, the principal needs to be able to use the information reported by the agents and recover the information that they actually received. To do so, the principal needs to be able to characterize the agents' behavior. The most direct way to address this problem would be to restrict the set of mechanisms used by the principal to mechanisms that are *incentive compatible* (IC) i.e., that incentivize agents to report their information truthfully. Indeed, when the mechanism chosen by the principal is non-IC, then it might be intractable for her to characterize agents' behavior, since she might need to find a NE of the n-players game induced by the principal, which is known to be PPAD-complete. Depending on whether the principal uses payment schemes or the action policy to induce particular players' behavior, the challenges that need to be solved to leverage the received feedback are different.

First, we analyze the case in which the principal does not have access to payments but can only influence agents' behavior by means of her action selection policy. In this context, we first show that it is impossible to learn an optimal mechanism while always being IC. Thus, we relax our requirement and ask that the mechanisms chosen by the principal are approximately IC on average. In particular, we require that the mechanisms chosen by the principal cumulatively violate the IC constraint only *sublinearly* w.r.t. the time horizon T. Following this relaxation, we study two distinct online learning scenarios in which the principal sequentially interacts with the agents. In the first one, which we call *full-feedback*, we assume that the feedback received by the principal is highly informative and independent from her mechanism. In this scenario, we were able to formulate an algorithm that achieves $\mathcal{O}(\sqrt{T})$ regret with respect to the optimal mechanism and $\mathcal{O}(\sqrt{T})$ cumulative violation of the IC constraint. Due to known lower bounds for multi-armed bandits, we can conclude that our algorithm is tight on its dependency from T. The second online learning scenario that we study is the *bandit-feedback* setting, in which we make minimal assumptions on the principal's knowledge and assume the least informative feedback possible. Here we were able to obtain an algorithm achieving $\mathcal{O}(T^{2/3})$ regret and cumulative IC violation, corroborated with a lower bound showing that our algorithm is tight on its dependency from T.

The second information acquisition case that we study is the one in which the principal is able to issue payments to the agents and the action chosen by her does not affect the agents' utilities. Such a setting is different from the previous one, since the principal can easily decode agents' feedback by choosing mechanisms that are *uncorrelated*, i.e., such that the payment received by an agent does not depend from the behavior of the others. However, we show that, in general, uncorrelated mechanisms are suboptimal. Thus, the principal should focus her attention on correlated mechanisms, but, similar to the previous case, this introduces complexities since it is not possible to always output correlated mechanisms that are IC. This suggests the need for a two-phase algorithm. In particular, we formulate an algorithm that uses the interpretability advantages of uncorrelated mechanisms during a first *estimation phase*, and then uses the estimates collected during this first phase to commit to an approximately optimal mechanism. Our algorithm achieves a $\mathcal{O}(T^{2/3})$ regret with respect to the optimal mechanism, which, based on lower bounds known for similar single-agent problems, constitutes a tight dependency from T.

The results presented in this section have been published as [3, 4].

References

1. Aumann, R.J.: Subjectivity and correlation in randomized strategies. J. Math. Econ. **1**(1), 67–96 (1974)
2. Brown, N., Sandholm, T.: Superhuman AI for multiplayer poker. Science **365**(6456), 885–890 (2019)
3. Cacciamani, F., Castiglioni, M., Gatti, N.: Online mechanism design for information acquisition. In: International Conference on Machine Learning. PMLR (2023)
4. Cacciamani, F., Castiglioni,M., Gatti,N.: Online information acquisition: hiring multiple agents. ICLR (2024)
5. Cacciamani, F., Celli, A., Ciccone, M., Gatti, N.: Multi-agent coordination in adversarial environments through signal mediated strategies (2021)
6. Campbell, M., Hoane, A.J., Jr., Hsu, F.-H.: Deep blue. Artif. Intell. **134**(1–2), 57–83 (2002)
7. Carminati, L., Cacciamani, F., Ciccone, M., Gatti, N.: A marriage between adversarial team games and 2-player games: enabling abstractions, no-regret learning, and subgame solving. In: International Conference on Machine Learning, pp. 2638–2657. PMLR (2022)
8. Celli, A., Gatti, N.: Computational results for extensive-form adversarial team games. In: Proceedings of the AAAI Conference on Artificial Intelligence, vol. 32 (2018)
9. Farina, G., Bianchi, T., Sandholm, T.: Coarse correlation in extensive-form games. In: Proceedings of the AAAI Conference on Artificial Intelligence, vol. 34, pp. 1934–1941 (2020)
10. Farina, G., Celli, A., Gatti, N., Sandholm, T.: Ex ante coordination and collusion in zero-sum multi-player extensive-form games. Adv. Neural Inf. Process. Syst. **31** (2018)
11. Farina, G., Celli, A., Marchesi, A., Gatti, N.: Simple uncoupled no-regret learning dynamics for extensive-form correlated equilibrium. J. ACM **69**(6), 1–41 (2022)
12. Lanctot, M., Waugh, K., Zinkevich, M., Bowling, M.: Monte Carlo sampling for regret minimization in extensive games. Adv. Neural Inf. Process. Syst. **22** (2009)
13. Moravčík, M., Schmid, M., Burch, N., Lisý, V., Morrill, D., Bard, N., Davis, T., Waugh, K., Johanson, M., Bowling, M.: Deepstack: Expert-level artificial intelligence in heads-up no-limit poker. Science **356**(6337), 508–513 (2017)
14. Moravcik, M., Schmid,M., Ha, K., Hladik, M., Gaukrodger, S.: Refining subgames in large imperfect information games. In: Proceedings of the AAAI Conference on Artificial Intelligence, vol. 30 (2016)
15. Nash, J.: Non-cooperative games. Ann. Math., 286–295 (1951)

16. Silver, D., Huang, A., Maddison, C.J., Guez, A., Sifre, L., Van Den Driessche, G., Schrittwieser, J., Antonoglou, I., Panneershelvam, V., Lanctot, M., et al.: Mastering the game of go with deep neural networks and tree search. Nature, **529**(7587), 484–489 (2016)
17. Vinyals, O., Babuschkin, I., Czarnecki, W.M., Mathieu, M., Dudzik, A., Chung, J., Choi, D.H., Powell, R., Ewalds, T., Georgiev, P., et al.: Grandmaster level in starcraft ii using multi-agent reinforcement learning. Nature **575**(7782), 350–354 (2019)
18. Von Stengel, B., Forges, F.: Extensive-form correlated equilibrium: definition and computational complexity. Math. Oper. Res. **33**(4), 1002–1022 (2008)
19. Zhang, B., Carminati, L., Cacciamani, F., Farina, G., Olivieri, P., Gatti, N., Sandholm, T.: Subgame solving in adversarial team games. Adv. Neural Inf. Process. Syst. **35**, 26686–26697 (2022)
20. Zhang, B.H., Farina, G., Anagnostides, I., Cacciamani, F., McAleer, S.M., Haupt, A.A., Celli, A., Gatti, N., Conitzer, V., Sandholm, T.: Computing optimal equilibria and mechanisms via learning in zero-sum extensive-form games. Adv. Neural Inf. Process. Syst. (2023)
21. Zhang, B.H., Sandholm, T.: Team correlated equilibria in zero-sum extensive-form games via tree decompositions. In: Proceedings of the AAAI Conference on Artificial Intelligence, vol. 36, pp. 5252–5259 (2022)
22. Zinkevich, M., Johanson, M., Bowling, M., Piccione, C.: Regret minimization in games with incomplete information. Adv. Neural Inf. Process. Syst. **20** (2007)

Technology and Applications of Compiler-Based Precision Tuning

Daniele Cattaneo⊙

Abstract In many computer architectures, high-precision calculations are ineffi-
cient and power-hungry. As a result, oftentimes it is valuable to exploit the tradeoff
between precision and performance to better utilize the hardware in ways that otherwise wouldn't be possible. Precision tuning is the practice of taking advantage of
this tradeoff, and it is very labour-intensive for the programmer to perform manually. Hence there is an increasing interest for compiler-based autotuners which
however are still imperfect and hard to use in practice. The underlying issue is that
the analyses and transformations required for precision tuning do not have a sufficient level of generality to be applicable to most existing programs. We attempt to
improve the state-of-the-art in precision tuning compilers by tackling this aspect,
introducing a novel data type allocation methodology and approaches for handling
mathematical functions and non-single-threaded programming. We also demonstrate
the applicability of Precision Tuning to applications based on machine learning and
safety-critical systems.

1 Introduction

As the need for computing power constantly increases, there is also an increasing but
contrasting need to use this power in more energy-efficient ways. Indeed, the energy
consumption of the ICT sector as a whole remains significant, and it is projected
to continue to increase over time [7]. One way to achieve this efficiency goal is
to exploit *approximate computing* techniques. Amongst them, one that is gathering
considerable interest is *precision tuning*. The goal of precision tuning is to reduce
the precision of a program by changing the data types employed in it. This achieves
a precision-performance tradeoff that allows to better exploit computing hardware
in general, especially considering embedded systems but also GPGPUs and HPC
nodes.

D. Cattaneo (✉)
DEIB, Politecnico di Milano, Milan, Italy
e-mail: daniele.cattaneo@polimi.it

© The Author(s) 2025
S. Garatti (ed.), *Special Topics in Information Technology*,
PoliMI SpringerBriefs, https://doi.org/10.1007/978-3-031-80268-3_3

At the moment, however, precision tuning has a limited appeal, is due to the complexity of selecting and implementing lower precision data types [2, 8]. Collecting precision requirements demands extensive domain experience, and selecting lower precision data types and algorithms is non-trivial. Despite this, precision tuning is usually done manually, rather than automatically. Whether automatic precision tuning is generally possible for any application is still an open research question. To answer this question, the most viable approach is to attempt building a new precision tuning framework that should be as extensible as possible. Then, this framework can be used in order to experiment with different types of analyses and transformations of the code, exploring the design space of precision tuning tools and building upon the current state-of-the-art. The experimental process also needs to exploit both synthetic benchmarks and real-world application software, in order to fully explore the diversity of possible use-cases. To this end, as a basis for our research, we develop and exploit the TAFFO precision tuning tool [1, 3].

The architecture of TAFFO is state-of-the-art with respect to current knowledge of precision tuning. It is based on the LLVM compiler construction tool, and as a result is implemented as five distinct code analysis and transformation passes. The first (Initialization) pass scans the source code of the application in search of user-inserted annotations. These annotations provide contextual metadata, such as value ranges for external inputs to the application, and indications about which region of code needs to be transformed. Then the Value Range Analysis pass extrapolates the ranges of all variables in the program from the given annotation and known instruction semantics. The ranges are then used to determine error boundaries and new data type assignments in the Data Type Allocation pass. Finally, the Conversion pass modifies the program to apply the new types, and the Error Analysis pass verifies a-posteriori if the error constraints are satisfied. The viability of TAFFO for transforming real-world embedded application has been proven on activity classification and system control tasks [5, 6].

In this chapter we briefly discuss how TAFFO handles data type allocation with integer linear programming and mathematical function optimization, and also how it can be employed in real-time systems. These topics aim to illustrate how TAFFO improves upon the previous state-of-the-art in precision tuning, and show that we have moved closer to the goal of making this technique accessible for general use.

2   Optimization-Based Data Type Allocation

The most crucial part of precision tuning, as realized in TAFFO and in other such tools, is the data type allocation. To this end, different tools take different approaches, but most of them – such as Precimonious [9]—adopt a *trial and error* approach which requires multiple recompilations of the software being optimized before reaching convergence. However this kind of solution scales poorly when the solution space is very large, i.e. when the number of data types to be considered is large. This issue is avoided by exploiting a fully static analysis, which does not need the code to be recompiled and run. We propose one such static analysis for data type allocation, which consists in translating the program to an optimization problem, which is then solved by an off-the-shelf tool. Our implementation allows seamless support not only of floating point and fixed point data types, but of any other number representation that might be adopted in the future. The solver is informed of the performance characteristics of the target platform, in order to allow it to perform an informed decision that not only involves the round-off error.

2.1 Comparing Floating Point and Fixed Point

To optimize the accuracy of the transformed program, we need to compare between different alternative representations. Unfortunately, each of these representations have different metrics to assess their precision. In our approach we define a new measurement unit, that can be applied to all numerical representations, the IEBW.

Definition 1 (*IEBW of a number*) The *Integer Equivalent Bit Width* (IEBW) for a number $x \in \mathbb{R}$ expressed in a representation $\mathcal{R}$ where ε is the smallest number for which $\mathcal{R}(x + \varepsilon) \neq \mathcal{R}(x)$ or $\mathcal{R}(x - \varepsilon) \neq \mathcal{R}(x)$ is defined as: $IEBW_{\mathcal{R}}(x) = -\lfloor \log_2 \varepsilon \rfloor$.

Informally, $IEBW_{\mathcal{R}}(x)$ is the minimum number of fractional bits needed by a fixed point representation to represent x with the same precision as its original representation $\mathcal{R}$. The IEBW can also be computed for a variable in a program.

Definition 2 (*IEBW of a variable*) The IEBW of a program variable v exploiting representation $\mathcal{R}$ and which can take values in the interval $[l, u]$ is defined as: $IEBW_{\mathcal{R}}(v) = \max\{IEBW_{\mathcal{R}}(x) \mid x \in [l, u]\}$.

It is easy to take the generic definition of the IEBW shown above and produce specialized definitions for a given representation. We show the definitions for floating point and fixed point representations.

Definition 3 (*IEBW of a floating point representation*) Let $x \neq \pm 0$ be a finite number represented in a binary floating point representation with precision p and maximum exponent E. We define its IEBW as $IEBW_{float(p,E)}(x) = p - \hat{p} - e_v$, where $e_v = \min(\lfloor \log_2 x \rfloor, E)$ is the exponent of x and $\hat{p}$ is 1 if $x \leq 2^{-E+1}$, and 0 otherwise.

Definition 4 (*IEBW of a fixed point representation*) Let x be a number in an unsigned fixed point representation of width $w > 0$, fractional bits $f < w$, and $w - f$ integer bits. Then, its IEBW is defined as $IEBW_{fix(w,f)}(x) = f$.

2.2   Model and Software Architecture

Our data type allocator exploits Integer Linear Programming (ILP) concepts to describe the precision-performance tradeoff in the program being optimized. Architecturally, it simply replaces the existing data type allocation in TAFFO, without additional modifications to its pipeline. In the model, the precision of each computation is defined in terms of the IEBW metric, by computing it from the value ranges provided by the Value Range Analysis. Additional parameters include target-hardware-specific characterization data, which includes mathematical operations as well as type casts, collected through micro-benchmarks.

The constraints in the model reflect the LLVM-IR representation of the program. The data type assigned to any floating point register a is modeled by a binary variable $x_{a,t}$ for each supported data type t. E.g., if in the solution $x_{a,binary32} = 1$, then the binary32 floating point format will be assigned to variable a. Only one of these variables must be active at once, which is enforced by constraints such as $\sum_{t \in \mathcal{T}} x_{a,t} = 1$. For any fixed point format $f \in \mathcal{T}_{fix}$ an additional integer variable $z_{a,f}$ contains the amount of fractional bits for a, if type f is chosen for it. More constraints are defined to model the peculiarities of LLVM-IR, such as type rules. The objective function takes into account the execution time of each operation and cast for each possible type, and the overall precision, expressed as the sum of the IEBW of all virtual registers. The terms related to execution time and cast are weighted separately to allow the user to fine-tune the required precision.

2.3   Experimental Comparison

We employed the benchmarks from the PolyBench/C test suite, version 4.2.1. This benchmark suite consists of several programs written in C that implement a wide variety of kernels used in various applications. No modifications were made to the benchmarks, except for the addition of the annotations required by TAFFO. In our evaluation, we employ four different machines with different hardware architectures: an STM3220G-EVAL evaluation board equipped with a 120 MHz CortexM3 ARM processor, a Raspberry Pi Model B Rev 2 single board computer featuring a BCM2835 system-on-a-chip (SOC) with an ARMv6 CPU core running at 800 MHz, a desktop computer featuring an Intel Pentium E5300 processor running at 2.6 GHz, and finally a NUMA node with four AMD Opteron 8435 CPUs, based on the K10 microarchitecture, running at 2.6 GHz.

On all architectures, the variation of the model tunables is correctly reflected in the experimental data for the great majority of benchmarks employed. By selecting a configuration that privileges performance, we achieve speedups as high as 800%—for the *heat3d* benchmark on the STM3220G-EVAL board. On the contrary, privileging error tolerance almost always produces programs that are functionally equivalent to non-tuned programs. While the gradual increase of the error is similarly reflected on all machines, the speedups are lower for the Intel and AMD machines, and occasionally a slowdown is obtained. We believe that this discrepancy occurs because our ILP model is based on a uniform cost model of each instruction. This kind of model does not accurately represent variable costs due to caching, pipelining, and other complex behavior exhibited by superscalar processors. The execution time increase directly or indirectly caused by the process of building and solving the ILP model is fairly small. The minimum compilation slowdown we measured across all compilations performed was $1.48\times$ (corresponding to an increase of the compilation time from 0.97 to 1.45 s), and the maximum slowdown was $3.25\times$ (corresponding to an increase from 0.66 to 2.16 s). The average slowdown was $2.10\times$. Thus, the overall slowdown can be expected to be in the order of $1.5 - 3\times$.

3 Handling Mathematical Functions

As we have seen, many computational kernels can gain considerable speedups from application of mixed-precision techniques. However, this speedup is often limited if the kernel makes heavy use of external library functions for performing operations such as the exponent, the square root, or computation of trigonometrical functions such as sine and cosine. This is due to the fact that often there are no readily available alternative implementations of these mathematical procedures for reduced-precision types, and as a result higher-precision versions are employed instead. Sometimes this does not affect the execution time, especially in the case in which we can employ hardware implementations made available by the CPU vendor. But when we need to fallback to software implementations, the lack of reduced precision has a measurable impact, especially in embedded systems.

Ideally, a precision tuning solution can rely on multiple implementations of each mathematical function, tailored to each supported data type. When the amount of such types is limited, this is not an issue. However, if the precision tuning tool supports configurable representations like fixed point ones, the amount of functions to develop and to manually provide grows substantially. In the context of programming languages, this kind of issue is solved using *template metaprogramming*, i.e. the ability to specify a generic version of a specific function or method, which then will be customized automatically during compilation depending on the way it is being used. Our proposal is to adapt the *metaprogramming* approach to the automatic generation of the implementation of mathematical functions when exploiting the fixed point representation, rather than providing every possible function ahead-of-time. This approach, which we call FIXM, allows to generate and handle only a restricted amount of functions, each one corresponding to the fixed point representations actually in use.

3.1 The Mathematical Function Generator

The implementation of FIXM is largely delegated to a new module for the TAFFO Conversion pass, named *Mathematical Function Generator*. This module detects where the program uses one of the supported POSIX standard mathematical functions, and generates the specialized fixed point mathematical functions for each call site. The original function calls in the program are then replaced with a call to the newly generated mathematical function. If a suitable function had already been generated, the existing function is used, instead of generating a new one. The process of generating the functions is performed by changing parameters in a set of template functions internally stored as LLVM-IR code. These parameters include the aforementioned function argument and return values.

Our generator can support two kinds of mathematical library implementation: CORDIC-based and Look-Up-Table-based. These different approaches have specific strong points and drawbacks that complement each other, giving a more complete set of options to the user of TAFFO. CORDIC provides a general method for computing trigonometric functions or other transcendental functions employing only additions and bit shifts. On the other hand, the Look-Up-Table (LUT) implementation exploits a pre-computed table of results for each mathematical function. Since storing the function's value for each possible input may take up too much memory, only values for evenly-spaced inputs are stored. When possible, the size of the LUT is reduced by exploiting periodicity and symmetry properties. While CORDIC produces more precise results, it is generally slower than the LUT approach.

In order to allow the automatic trade-off of code size and execution time, we added a parameter to FIXM called Z which expresses the proportion of space available for additional code, excluding any occupation attributable by the optimizations performed by FIXM. Depending on the value of Z, the Mathematical Function Generator decides at compile time whether to generate a look-up-table or a CORDIC implementation of a given trigonometric function. By default, Z is set to 0, and in that setting it forces the Mathematical Function Generator to always generate CORDIC implementations. Conversely, $Z = 1$ will always generate LUTs instead of using CORDIC. The Mathematical Function Generator models such decision process as a knapsack problem, whose solution is obtained using the greedy algorithm proposed by Dantzig [4].

3.2 Evaluation of FIXM

We demonstrate the capabilities of FIXM by exploiting two benchmarks from the AXBENCH [11] suite: *InverseK2J* and *FFT*. Additionally, we evaluated FIXM on *FBench* [10], a synthetic benchmark for floating point performance. These three benchmarks are representative of real-world applications exploiting complex mathematical functions such as sine and cosine. The evaluation is performed on two microcontroller boards: the same STM3220G-EVAL as described in Sect. 2.3, and a STM32F4Discovery board with an ARM Cortex M4 CPU running at 168 MHz. The execution times are compared with the same benchmark compiled with TAFFO but without FIxM.

We find that FIxM brings a substantial benefit in terms of execution time independently from the hardware class. In fact, we see high speedups for both boards, with the STM3220G-EVAL board taking the lead due to its use of the Cortex M3 CPU which lacks floating point support entirely. The FFT benchmark achieves the highest speedup of 181% on this board. The speedup for InverseK2J is 72% and for FBench it is 31%. The speedups for the STM32F4Discovery board are slightly lower: 125%, 60% and 19% respectively. The entirety of the speedup we observe is attributable to the FIxM approach. In fact, without FIxM, using TAFFO on these benchmarks does not bring any significant speedup improvement. For all benchmarks the average relative error figures are under 5×10^{-4}.

4 WCET Optimization

Finally, we consider an application scenario where precision tuning may be applied: *critical* and *mixed critical* systems. In the development of such systems, one important aspect is to analyze the *Worst-Case Execution Time* (WCET) of each routine involved in the critical task. Knowing the WCET allows to schedule multiple tasks such that the CPU time allotted for each one of them will be appropriate. Henceforth, in many applications there is no single WCET target to hit, and the WCET is a metric to be optimized. This optimization can be performed through precision tuning as well.

The basic idea is to adapt the Integer Linear Programming optimizer described in Sect. 2 to make an estimation of the WCET. Indeed, the mathematical model already computes a parametric estimation of the relative execution time slowdown and of the round-off error, with respect to a fully floating-point-based implementation. These estimates are not useful in general to gauge the real error or execution time of the generated program, because the per-instruction coefficients employed are ratios rather than absolute errors or instruction timings. However, in principle, if we replace such coefficients with values corresponding to the actual error being inserted by a computation or its actual execution time in clock cycles, the optimiser will gain visibility to a realistic estimation of these metrics. This enables optimising the program for a given maximum error or execution time.

The combination of automatic WCET estimation and the ability to control it via compiler optimization allows to not only build a WCET-aware compiler but also to target a specific WCET. As a result, this work enables a tradeoff between precision and performance also in the field of real-time and critical systems. We believe this combination to be novel in the field of WCET optimization at large.

4.1 Implementation

Let us consider the ILP model constructed in Sect. 2. Previously we did not make any distinction between each basic block, however to produce a realistic estimation of the execution time we must properly weigh the *time* component relative to each basic block in order to properly represent conditionals and loops. On an in-order CPU architecture such as a microcontroller, the execution time of a basic block can be modelled with good accuracy as the sum of the individual execution times of each instruction in the basic block. With this assumption, we can adjust the existing model used for precision tuning alone to represent the entire program by adding constant terms for instructions that are not part of the precision tuning optimization. The error terms in the model stay unchanged. The newly reformulated execution time terms can now be employed in a linear constraint which statically imposes a limit on the worst-case execution time (WCET) of the program.

4.2 Analysis of the Methodology

We performed a set of experiments aimed at testing the quality of the execution time estimation. As example applications, we chose some benchmarks from the PolyBench/C suite. Of the entire set of benchmarks, we chose the ones with the highest execution time variance depending on the optimisation parameters: *2mm*, *3mm*, *covariance*, *lu* and *nussinov*. The benchmarks are unmodified, exception done for the addition of the required annotations for TAFFO. Hardware-wise, the platform targeted for the estimation was the STM3220G-EVAL ST Microelectronics embedded evaluation board, with a 120 MHz Cortex-M3 ARM processor. The experiments evidence that the execution time prediction is consistent with the measured execution time: speedups happen exactly when they are predicted by the model. This evidences that the execution time estimation performed by the ILP model is usable as a proxy for the real execution time.

5 Conclusion

We have discussed precision tuning—the practice of enacting a performance-precision tradeoff—within the context of the TAFFO tool. Given this premise we described novel techniques that makes precision tuning more applicable to real-world applications: a new data type allocation approach based on Integer Linear Programming, and a metaprogramming-inspired approach to automatically generate customized mathematical function implementations. Exploiting our methodologies, we achieved positive results overall: in particular, speedups up to $5\times$ and more, with error figures that are overall under 1%. Finally, we describe a way to exploit the program model we developed for the data type allocation to predict the Worst-Case Execution Time required by a software routine, which allows us to perform just enough optimizations to ensure it meets real-time requirements.

This work is just a step towards the ultimate goal of *plug-and-play precision tuning*, i.e. the capability of enabling precision tuning at the turn of a switch. Further improvements to TAFFO as a tool will therefore be necessary to reach that ideal. One addition could be the implementation of dynamic analyses of the range and of the error, in alternative to the static approach if its results are too conservative. Another important improvement we envision is the support for *high-level synthesis* (or HLS) backends for the automatic synthesis into hardware of a program. These future developments can also be compounded by the adoption of next-generation compiler construction frameworks such as MLIR, which will allow precision tuning to be used in an even wider range of contexts and applications.

Acknowledgements I would like to thank Giovanni Agosta, Stefano Cherubin, Michele Chiari, Nicola Fossati and Gabriele Magnani for their contributions to the work presented in this chapter.

References

1. Cattaneo, D., Chiari, M., Agosta, G., Cherubin, S.: Taffo: the compiler-based precision tuner. SoftwareX **20**, 101238 (2022)
2. Cherubin, S., Agosta, G.: Tools for reduced precision computation: a survey. ACM Comput. Surv. **53**(2) (2020)
3. Cherubin, Stefano, Cattaneo, Daniele, Chiari, Michele, Di Bello, Antonio, Agosta, Giovanni: TAFFO: tuning assistant for floating to fixed point optimization. IEEE Embed. Syst. Lett. **12**(1), 5–8 (2020)
4. Dantzig, G.B.: Discrete-variable extremum problems. Oper. Res. **5**(2), 266–288 (1957)
5. Fossati, N., Cattaneo, D., Chiari, M., Cherubin, S., Agosta, G.: Automated precision tuning in activity classification systems: a case study. In: Proceedings of the 11th Workshop on Parallel Programming and Run-Time Management Techniques for Many-core Architectures and the 9th Workshop on Design Tools and Architectures For Multicore Embedded Computing Platforms, PARMA-DITAM '20, pp. 1–6 (Jan. 2020)
6. Magnani, G., Cattaneo, D., Chiari, M., Agosta, G.: The impact of precision tuning on embedded systems performance: a case study on field-oriented control. In: Bispo, J., Cherubin, S., Flich, J. (eds.) 12th Workshop on Parallel Programming and Run-Time Management Techniques for Many-core Architectures and 10th Workshop on Design Tools and Architectures for Multicore Embedded Computing Platforms (PARMA-DITAM 2021), vol. 88 of Open Access Series in Informatics (OASIcs), pp. 3:1–3:13. Dagstuhl, Germany, Schloss Dagstuhl – Leibniz-Zentrum für Informatik (2021)
7. Masanet, Eric, Shehabi, Arman, Lei, Nuoa, Smith, Sarah, Koomey, Jonathan: Recalibrating global data center energy-use estimates. Science **367**(6481), 984–986 (2020)
8. Mittal, S.: A survey of techniques for approximate computing. ACM Comput. Surv. **48**(4), 62:1–62:33 (2016)
9. Rubio-González, C., Nguyen, C., Nguyen, H.D., Demmel, J., Kahan, W., Sen, K., Bailey, D.H., Iancu, C., Hough, D.: Precimonious: tuning assistant for floating-point precision. In: Proceedings of the International Conference on High Performance Computing, Networking, Storage and Analysis, SC '13, pp. 27:1–27:12 (Nov. 2013)

10. Walker, J.: Floating point benchmarks (2016)
11. Yazdanbakhsh, Amir, Mahajan, Divya, Esmaeilzadeh, Hadi, Lotfi-Kamran, Pejman: AxBench: a multiplatform benchmark suite for approximate computing. IEEE Des. Test **34**(2), 60–68 (2017)

Resource Allocation and Scheduling Problems in Computing Continua for Artificial Intelligence Applications

Federica Filippini ⓘ

Abstract The problem of optimizing the execution of Artificial Intelligence (AI) and Deep Learning (DL) applications in the Computing Continuum gained remarkable popularity in recent years, due to both the widespread adoption of AI in real-life scenarios and the challenging environment introduced by a distributed Edge-to-Cloud paradigm. We tackled the resource selection, scheduling and placement problem both from a design-time and runtime perspective, considering, on one hand, AI inference applications characterized by complex workflows with multiple heterogeneous components and, on the other hand, resource-demanding DL training jobs executed on public or private GPU-accelerated clusters.

1 Introduction

These years witness two interconnected phenomena: on one hand, the increasing pervasiveness of Deep Learning (DL) and Artificial Intelligence (AI), widely adopted in largely heterogeneous fields (e.g., personalized healthcare, industrial applications, etc.); on the other hand, an accelerated migration towards distributed computing. Latency and privacy constraints limit the feasibility of processing in the public Cloud the large data volumes generated by Internet of Things devices, favouring on-premises resources or Edge devices. However, large datacenters featuring powerful, often GPU-accelerated, Virtual Machines (VMs) are still crucial for resource-demanding tasks, which benefit from the possibility of accessing ideally unlimited computational and storage resources according to pay-to-go pricing models. The effective management of such a complex environment, featuring heterogeneous Edge-to-Cloud resources interconnected in a so-called Computing Continuum, poses great challenges [6]. The execution of AI inference workflows including diverse components needs to be optimized at design time, selecting the most appropriate resources from the available computational layers, and then adapted at runtime, since fluctuating workload and environment conditions may determine sub-optimal scenarios

F. Filippini (✉)
Politecnico di Milano, Milan, Italy
e-mail: federica.filippini@polimi.it

© The Author(s) 2025
S. Garatti (ed.), *Special Topics in Information Technology*,
PoliMI SpringerBriefs, https://doi.org/10.1007/978-3-031-80268-3_4

where resources are saturated or under-utilized. Similarly, DL training applications executed in private or public Cloud clusters are to be effectively scheduled on GPU-accelerated nodes to minimize the expected costs and meet the user-imposed due dates. This work addresses these three problems from different perspectives, proposing mathematical models and heuristic or Reinforcement Learning (RL)-based methods for the efficient and effective management of AI inference applications, both at design time and runtime, and of DL training jobs. Furthermore, it discusses techniques to develop analytical and Machine Learning (ML)-based performance models, crucial to accurately predict the applications response times on heterogeneous resources. The proposed tools are validated in both simulated and prototype environments, proving their effectiveness and applicability to practical scenarios.

In the following, Sect. 2 overviews our work on performance modeling based on ML; Sect. 3 presents the framework we propose for the design-time optimization of AI workflows. Sections 4 and 5 address the runtime resource management in the Computing Continuum, focusing on AI inference applications and DL training jobs, respectively. Finally, Sect. 6 draws conclusions and ideas for future works.

2 Performance Models

Predicting the execution times of computing tasks (e.g., a DL training job or a stage of a complex inference pipeline) executed on different resources is crucial to support their optimal placement, since accurate estimates allow to correctly forecast their computational requirements and the expected costs related to the Computing Continuum resources usage [17]. We addressed this problem both through analytical, white-box methods based on queuing theory, which guarantee a fast computation and the possibility of considering components interference due to co-location, and by developing black-box models based on ML. Once trained on profiling data, these allow to predict the response times of tasks in unforeseen conditions, interpolating or extrapolating over the available features, when information about the demanding times or the components throughput are not easily available.

In the next sections, we firstly present OSCAR-P [13], a tool that seamlessly supports the execution, profiling and performance modeling for serverless AI workflows running in the Computing Continuum. Then, we discuss the development of ML-based models for Distributed Deep Learning (DDL) training applications [9].

2.1 OSCAR-P: Serverless Applications Performance

OSCAR-P is an auto-profiling tool we built around OSCAR [21], a state-of-the-art, open-source platform that supports the execution of serverless applications in Cloud and Edge environments. OSCAR-P allows to automatically test the application under study on heterogeneous resources and in different hardware configurations, gathering information on the components response times. Individual ML-based performance models for each component-resource pair are built through *aMLLibrary*, an open-source tool for the automatic generation of ML-based regression models [15]. They are combined to predict the response time of the entire workflows, thus supporting both their design-time optimization [23] and runtime management [18].

The models generated during the respective experimental validation, which considered several multi-component AI workflows deployed in the Computing Continuum, achieved good performance, with a Mean Absolute Percentage Error (MAPE) lower than 10% on interpolation and 20% on extrapolation in the analyzed scenarios.

2.2 *Distributed Deep Learning on Ray*

In addition to the performance of AI inference pipelines, we worked on estimating the execution time of DDL training applications, focusing on jobs developed and executed on Ray clusters. Ray [20] is an increasingly popular open-source framework that supports the scaling of general-purpose and AI workflows. The DDL paradigm [4] was developed to accelerate the training of DL jobs by subdividing it in multiple tasks executed in parallel. To the best of our knowledge, no previous studies on the performance of DDL applications executed on Ray clusters are available.

We considered two Convolutional Neural Network (CNN) models for image classification as target training jobs. The limited size of the target networks allowed to evaluate both GPU and CPU scalability; moreover, the considered scenario is particularly challenging for the development of accurate ML-based performance models, since the profiling data collected during the training are more noisy due to the communications overhead. Jobs were deployed on a cluster including several worker nodes hosted on AWS [1]. We profiled jobs collecting the execution time of several stages (from a single training epoch to the loop including inter-workers communications) and exploited aMLLibrary to train ML models using as base features the number of used resources and the dataset size. Our models perform well for both interpolation and extrapolation, achieving a MAPE between 3 and 15%.

3 Design-Time Optimization

The deployment and execution of AI applications in the Computing Continuum is challenging, and careful Resource Selection and Component Placement (RS-CP) decisions are crucial to find a mapping between the applications components and the infrastructure that minimizes the expected costs while meeting hardware, network, privacy and QoS requirements [6] (see Fig. 1). At design time, according to the expected value of input workload, application owners need to: (i) select the best resource among the available alternatives in each computational layer; (ii) identify the optimal deployment for Deep Neural Network (DNN) components that

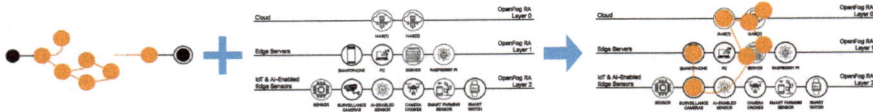

Fig. 1 The placement problem: mapping an application on the infrastructure

can be partitioned at different layers [16] to facilitate offloading; (iii) determine the minimum-cost placement of all component partitions on the selected resources, under the users-imposed constraints. A good-quality RS-CP solution is crucial also to facilitate runtime management tools that adapt the design-time choices to varying workload conditions (see Sect. 4). Indeed, a wrong resource selection at design-time may prevent such methods from suitably identifying a feasible reconfiguration (particularly for Edge devices, which have limited capacity and usually become the performance bottleneck if the input workload increases).

To the best of our knowledge, our work represents one of the first attempts to effectively address the issue of resource contention, by adopting queuing theory to model application components response times [24], while existing design-time tools usually focus on the execution of a single application instance. Similarly, considering multiple candidate DNN partitioning points for our application components and identifying the optimal deployment as part of the design-time optimization represents a novel contribution with respect to the existing literature [23].

We proposed a Mixed-Integer Non-Linear Programming (MINLP) formulation for the design-time RS-CP problem in the Computing Continuum, with objective:

$$\min \left(C_E + C_C + C_F + C_T \right), \tag{1}$$

where C_E is the cost of Edge devices, including the amortized investment cost and the expected management costs over the single run of the target application; C_C is the usage cost of Cloud VMs, which depends on the chosen provider; C_F denotes the cost of FaaS instances, which depends on the memory size, the functions duration and the total number of invocations; C_T is a state transition cost that some serverless providers charge for the message passing and coordination between successive functions. This is subject to hardware, memory, and QoS constraints imposed on single application components or complex execution paths, which make the problem $\mathcal{NP}$-hard.

We developed SPACE4AI-D (System Performance and Cost Evaluation on Cloud for AI applications Design), a tool that leverages Random Search and several heuristic methods (i.e., Local Search, Tabu Search, Simulated Annealing and Genetic Algorithms), to determine good-quality solutions in a reasonable time. SPACE4AI-D yields significant cost reductions with respect to the state of the art, between 27.6% and 58% on average against a tool proposed in [19], and the observed deviation between real and predicted costs in a prototype environment is lower than 13%.

4 Runtime Management for Inference Applications

In practical applications, the design-time optimization discussed in Sect. 3 has to be continuously reevaluated and, possibly, updated: the expected application workload is often subject to fluctuations due to, e.g., variations in the generated data volumes [23]. Therefore, the initial placement may become unfeasible or lead to resource under-utilization, dictating the need of effective monitoring and adaptation mechanisms. The most relevant difference between design-time and runtime optimization methods is that, while a design-time tool is allowed to take as much time as needed (up to several minutes) to find the initial production deployment (i.e., the initial placement), a runtime tool must provide a feasible reconfiguration in few seconds at most, due to the online running application. Therefore, runtime tools must be designed according to appropriate cost-reactivity trade-offs while tuning the solver algorithms, as well as implemented in a very efficient way to reduce the execution time. Among the families of approaches that can be leveraged to develop runtime optimization frameworks, we focused on heuristic-based methods (see Sect. 4.1) and RL-based methods (see Sect. 4.2). The former guarantee good solution quality and fast execution times, but they require a deep knowledge of the problem characteristics and sometimes restrictive assumptions on the model properties to guarantee effective implementations. On the other hand, RL-based methods can successfully adapt overtime to variations in, e.g., the workload model or the application computing requirements by relying only on observations extracted from the environment [3]. Therefore, they are often more flexible than methods based on analytical optimization models, but usually require a long time to learn an effective policy. During the training, an agent needs to explore all the possible actions, possibly incurring in severe constraints violations or overspending. Our work aims to bring together the best of the two worlds: the RL-based framework we propose accelerates the initial training by leveraging the prior knowledge coming from the design-time analysis.

4.1 SPACE4AI-R

SPACE4AI-R (System Performance and Cost Evaluation on Cloud for AI applications Runtime) is the runtime counterpart of SPACE4AI-D, and exploits an efficient Random Search combined with a Stochastic Local Search algorithm to identify a suitable reconfiguration of the running placement coping with the workload fluctuations [10]. In particular, reconfiguration decisions cannot change the selected resource type on computational layers of physical resources, while they include: (i) identifying which resource is best at each virtual and currently unused computational layer or at the FaaS layer, (ii) scaling in/out resources at the physical layers or at currently-used virtual layers, (iii) selecting the optimal DNN deployment for each component, and (iv) assigning each partition to a compatible resource.

The experimental validation proved the advantages of a dynamic system reconfiguration over a static approach where the design-time solution is kept fixed regardless the variations in the input workload, achieving an average cost reduction between 2 and 36% in the analyzed scenarios. Moreover, SPACE4AI-R yields a speed-up close to $100\times$ with respect to its design-time counterpart, tackling problem instances involving up to 15 AI application components in less than 1.5 s on average. Finally, the comparison with a state-of-the-art method that performs scaling actions with the goal of keeping the resources utilization within a predefined interval [26] highlights cost reductions between 10% and 40% and a better capability of avoiding response-time constraints violations.

4.2   FIGARO

FIGARO (reinForcement learnInG mAnagement acRoss the computing cOntinuum) is an RL-based approach that automatically adapts the current deployment to varying and possibly unforeseen conditions [7]. To limit the negative impact of the initial exploratory phase, FIGARO undergoes an initial training that exploits imitation learning to develop a policy close to the one exploited by SPACE4AI-R. This is further trained via interactions with a simulated environment, leveraging a Deep Q-Learning approach to develop a good-quality scaling policy.

A preliminary evaluation of the FIGARO framework proved the effectiveness of our RL Agent in a simplified scenario involving only scaling actions. In particular, the initial offline training stage that exploits the design-time knowledge reduces the cumulative number and entity of response time constraints violations by 2–4 orders of magnitude, and by 25% the number of required training iterations.

5   Runtime Management for Training Applications

Training DL applications is a huge challenge these days. Jobs are very resource-demanding, despite the significant performance benefits guaranteed by GPU acceleration [25]. Moreover, high-performance servers are characterized by considerable costs, both to acquire and maintain resources (due to, e.g., energy consumption and cooling [5]) and in the public Cloud. Consequently, effectively tackling the Resource Selection and Job Scheduling (RS-JS) problem is crucial both for developers that rely on public services and for Cloud Service Providers (CSP). One one hand, selecting the most suitable resources to co-locate different DL training workloads and determining efficient schedules allows Cloud users to minimize the execution costs of their applications (see Sect. 5.1). On the other hand, CSPs need efficient methods to select resources for the execution of each job to minimize the energy consumption costs while meeting the applications due dates (see Sect. 5.2).

5.1 RS-JS Problem from a User's Perspective

We envisioned a scenario including a multi-node cluster, where servers can be provisioned on-demand and configured with heterogeneous VMs characterized by a possibly different number of GPUs, according to the provider catalog (see, e.g., [1]). The submitted jobs can be executed concurrently on the assigned nodes; all computational resources are shared except for GPUs, which are dedicated to running single jobs. Determining the optimal solution means addressing two intertwined subproblems: (i) the RS problem consists in identifying, for each submitted job, the optimal VM type and number of GPUs to minimize its execution cost while meeting the prescribed due date; (ii) the JS problem involves determining which jobs to run if the available resources are not enough to execute all of them concurrently, and assigning them to the cluster nodes effectively partitioning the selected resources. We tackle these problems jointly and in an online setting, without any information about the number and characteristics of jobs that will be submitted in the future.

We developed a MILP formulation and a heuristic algorithm based on Randomized Greedy and Path Relinking [14]. The MILP objective reads:

$$\min \sum_{j} \left(C_j^D + C_j^P \right) + \sum_{jn} C_{jn}^E + \sum_{n} C_n^{idle}. \tag{2}$$

For each job j, we pay a penalty C_j^D if it exceeds the due date, and an additional cost C_j^P if its execution is postponed: while our scheduler supports pre-emption, this term incentivizes a prompt execution to minimize starvation. Moreover, we consider the execution cost C_{jn}^E paid if job j is deployed on node n, and a final penalty C_n^{idle} that, for each cluster node n, reduces the risk of leaving idle resources.

Our heuristic scheduler proceeds by sorting jobs according to the *pressure*, i.e., the distance from the due date, so that the most critical receive resources as first. The optimal configuration for each job is assumed to be: (i) the cheapest configuration such that it is executed before its due date, if possible, or (ii) the fastest available configuration if the due date cannot be met with any setup. Jobs-to-nodes assignments are finally performed according to a best-fit approach to limit idle resources.

The experimental validation compares Path Relinking, simpler heuristic variants based on Randomized Greedy, a previously-proposed method based on a hierarchical problem decomposition [8], and a dynamic programming-based method adapted from the literature [22]. Even in the most complex settings, we achieve an average cost reduction between 23 and 97% and tackle instances including up to 100 cluster nodes and 450 concurrent jobs in less than 7 s. The deviation between real and predicted costs in a prototype environment is lower than 5%.

5.2   RS-JS Problem from a Cloud Provider Perspective

In this scenario, we assume that jobs may conclude the training before the predefined number of epochs, since DL developers usually rely on termination criteria based on, e.g., the accuracy level; therefore, we consider stochastic execution profiles. Guided by a solution model proposed in [2], which identifies the optimal schedule for a single job exploiting information about the probability distribution of the epochs required to complete the training and the power consumption model, we proposed a MINLP formulation for the minimization of the expected energy costs and designed a STochastic Scheduler (STS) to select the best GPU type and amount of resources to execute each application, incorporating GPU sharing [11]. STS exploits the assumption that energy costs increase linearly with the number of used GPUs, which can be observed in practice from Cloud providers pricing models (see, e.g., [1]), and designs the resource profile for each job starting the execution with a low-power configuration, which is usually less expensive, and accelerating the training process by progressively increase the resources as the due date approaches.

STS outperforms both state-of-the-art approaches [22] and an adapted version of our Randomized Greedy algorithm [12], achieving cost reductions between 38 and 80% on average. Finally, the adoption of GPU sharing guarantees additional advantages, further reducing the STS costs by 19–29% on average.

6   Conclusions

The approaches proposed in this work focus on optimizing the execution of AI and DL applications in the Computing Continuum. We initially worked on the automatic development of ML-based models for performance prediction, and then tackled the resource selection, scheduling and placement problems both from a design-time and runtime perspective. The frameworks we developed target, on one hand, AI inference applications characterized by complex workflows with multiple heterogeneous components and, on the other hand, resource-demanding DL training jobs executed on public or private GPU-accelerated clusters.

All the proposed approaches proved their effectiveness and applicability to practical scenarios, achieving significant cost reductions against the state of the art with minimal deviations between real and predicted costs when evaluated in prototype environments. Future works will further investigate the potential of RL for the runtime applications management, exploring federated and multi-agent approaches.

References

1. Amazon EC2 Instance Types (2023). https://aws.amazon.com/ec2/instance-types. Accessed 11 Jul. 2024
2. Anselmi, J., Gaujal, B.: Energy optimal activation of processors for the execution of a single task with unknown size. In: MASCOTS, pp. 65–72 (2022). https://doi.org/10.1109/MASCOTS56607.2022.00017
3. Barrett, E., Howley, E., et al.: Applying reinforcement learning towards automating resource allocation and application scalability in the cloud. Concurr. Comput.: Pract. Exp. 25(12), 1656–1674 (2013). https://doi.org/10.1002/cpe.2864
4. Ben-Nun, T., Hoefler, T.: Demystifying parallel and distributed deep learning: an in-depth concurrency analysis. ACM Comput. Surv. 52(4) (2019). https://doi.org/10.1145/3320060
5. Bharany, S., Sharma, S., et al.: A systematic survey on energy-efficient techniques in sustainable cloud computing. Sustainability 14(10) (2022). https://doi.org/10.3390/su14106256
6. Costa, B., Bachiega, J., et al.: Orchestration in fog computing: a comprehensive survey. ACM Comput. Surv. 55(2) (2022). https://doi.org/10.1145/3486221
7. Filippini, F., Cavadini, R., et al.: FIGARO: reinForcement learnInG mAnagement acRoss computing continua. In: IEEE/ACM UCC Proceedings (2023). https://doi.org/10.1145/3603166.3632565
8. Filippini, F., Lattuada, M., et al.: Hierarchical scheduling in on-demand GPU-as-a-service systems. In: SYNASC, pp. 125–132. IEEE (2020). https://doi.org/10.1109/SYNASC51798.2020.00030
9. Filippini, F., Lublinsky, B., et al.: Performance models for distributed deep learning training jobs on ray. In: SEAA, pp. 30–35 (2023). https://doi.org/10.1109/SEAA60479.2023.00014
10. Filippini, F., Sedghani, H., Ardagna, D.: SPACE4AI-R: a runtime management tool for AI applications component placement and resource scaling in computing continua. In: IEEE/ACM UCC Proceedings (2023). https://doi.org/10.1145/3603166.3632560
11. Filippini, F., Anselmi, J., et al.: A stochastic approach for scheduling AI training jobs in GPU-based systems. IEEE Trans. Cloud Comput. 12(01), 53–69 (2024). https://doi.org/10.1109/TCC.2023.3336540
12. Filippini, F., Lattuada, M., et al.: A path relinking method for the joint online scheduling and capacity allocation of DL training workloads in GPU as a service systems. IEEE Trans. Serv. Comput. 16(3), 1630–1646 (2023). https://doi.org/10.1109/TSC.2022.3188440
13. Galimberti, E., Guindani, B., et al.: OSCAR-P and AMLLibrary: performance profiling and prediction of computing continua applications. In: ACM/SPEC ICPE Companion, pp. 139–146 (2023). https://doi.org/10.1145/3578245.3584941
14. Gendreau, M., Potvin, J.Y. (eds.): Handbook of Metaheuristics. International Series in Operations Research & Management Science. Springer International Publishing (2019)
15. Guindani, B., Lattuada, M., Ardagna, D.: aMLLibrary: an AutoML approach for performance prediction. In: ECMS Proceedings, pp. 215–221 (2023). https://doi.org/10.7148/2023-0215
16. Kang, Y., Hauswald, J., et al.: Neurosurgeon: collaborative intelligence between the cloud and mobile edge. In: ASPLOS Proceedings, pp. 615–629 (2017). https://doi.org/10.1145/3037697.3037698
17. Lattuada, M., Gianniti, E., et al.: Performance prediction of deep learning applications training in GPU as a service systems. Clust. Comput. 25(2), 1279–1302 (2022)
18. Li, E., Zeng, L., et al.: Edge AI: on-demand accelerating deep neural network inference via edge computing. IEEE Trans. Wirel. Commun. 19, 447–457 (2019)
19. Lin, C., Khazaei, H.: Modeling and optimization of performance and cost of serverless applications. IEEE Trans. Parallel Distrib. Syst. 32(3), 615–632 (2021)
20. Ray (2023). https://www.ray.io. Accessed 05 Jul. 2024
21. Risco, S., Moltó, G., et al.: Serverless workflows for containerised applications in the cloud continuum. J. Grid Comput. 19(3), 1–18 (2021)

22. Saxena, V., Jayaram, K.R., et al.: Effective elastic scaling of deep learning workloads. In: MASCOTS, pp. 1–8 (2020). https://doi.org/10.1109/MASCOTS50786.2020.9285954
23. Sedghani, H., Filippini, F., Ardagna, D.: A random greedy based design time tool for AI applications component placement and resource selection in computing continua. In: IEEE EDGE, pp. 32–40 (2021). https://doi.org/10.1109/EDGE53862.2021.00014
24. Sedghani, H., Filippini, F., Ardagna, D.: A randomized greedy method for AI applications component placement and resource selection in computing continua. In: IEEE JCC, pp. 65–70 (2021). https://doi.org/10.1109/JCC53141.2021.00022
25. Shawi, R.E., Wahab, A., et al.: DLBench: a comprehensive experimental evaluation of deep learning frameworks. Clust. Comput. **24**(3), 2017–2038 (2021)
26. Zhu, X., Young, D., et al.: 1000 islands: an integrated approach to resource management for virtualized data centers. Clust. Comput. **12**, 45–57 (2009)

Electronics

Analog Circuit Design for In-Memory Linear Algebra Accelerators

Piergiulio Mannocci and **Daniele Ielmini**

Abstract Since its introduction in 1945, computing systems have been built around von Neumann's architecture, predicating the physical separation of memory and computing units on grounds of flexibility and generality. However, the increasingly data-driven workloads of modern-day applications exacerbate the energy and latency overheads associated with data shuttling. In-memory computing (IMC) radically subverts the classical paradigm by performing computation in situ within the memory elements, unlocking theoretically unrivaled throughput and energy efficiency. Among the wide spectrum of IMC architectures, closed-loop in-memory computing (CL-IMC) has attracted interest for its capability to accelerate computationally heavy operations of increasing use, such as matrix inversion. This chapter focuses on analog closed-loop circuits for in-memory accelerators. A mathematical framework is derived to develop a matrix-based circuit simulator providing orders-of-magnitude speedups with respect to SPICE solvers. New circuits for the acceleration of regularized regressions and linear quadratic estimation are characterized in terms of accuracy and speed, providing improvement with respect to digital solvers in baseband processing in 6G systems and Kalman filters. Experimental demonstrations finally provide a real-world implementation of CL-IMC topologies. The obtained results strengthen the position of CL-IMC as a promising candidate for next-generation energy-efficient algebraic accelerators.

1 Introduction

Modern computers are still designed around the von Neumann architecture [1], where processing and memory units are physically separated as shown in Fig. 1a. Data are fetched from the memory unit, shuttled to the processing unit (where computation

P. Mannocci (✉) · D. Ielmini
Dipartimento di Elettronica, Informazione e Bioingegneria, Politecnico di Milano, Milano, Italy
e-mail: piergiulio.mannocci@polimi.it

D. Ielmini
e-mail: daniele.ielmini@polimi.it

© The Author(s) 2025
S. Garatti (ed.), *Special Topics in Information Technology*,
PoliMI SpringerBriefs, https://doi.org/10.1007/978-3-031-80268-3_5

45

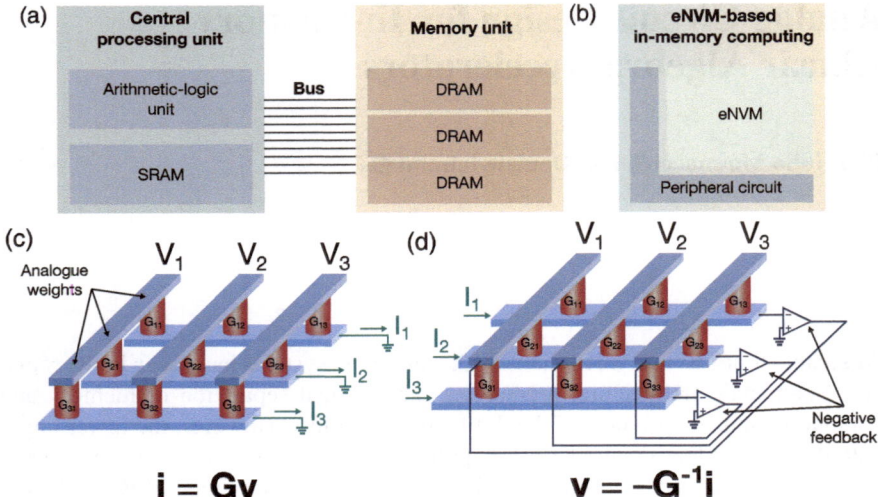

$$i = Gv \qquad\qquad v = -G^{-1}i$$

Fig. 1 Memory-processing integration. **a** Prototypical von Neumann architecture, where memory and processing units are physically separated and interconnected by a data bus. **b** In-memory computing, where computation happens directly within the embedded nonvolatile memory array. **c** Open-loop crosspoint-based primitive for matrix-vector multiplication. **d** Closed-loop crosspoint-based primitive for inverse-matrix-vector multiplication. **a** and **b** Reproduced with permission from [3]. Copyright 2023 Author(s), licensed under a Creative Commons Attribution 4.0 License

occurs), and then returned to memory for storage. Nowadays, as problems require more and more data to be analyzed, most of the time and energy is spent on handling this continuous data transfer [2], an issue known as *von Neumann bottleneck*. Moreover, this scheme is intrinsically limited by the rate at which data can be transferred between the two neighboring units, a parameter hindered by the technological development difference between the computing and memory units. The latter is generally slower than the former, an issue termed *memory wall*.

The energy and latency overheads associated with the constant data transfer can be suppressed by performing processing in situ within the memory, which is the core concept of in-memory computing (IMC) [3]. Figure 1b shows a prototypical IMC core, where an embedded nonvolatile memory (eNVM) is complemented with dedicated peripheral circuitry, typically including converters and access decoders. Emerging memory devices are the most suitable candidate to enable eNVM-based IMC thanks to their nonvolatility and high scalability potential. At the same time, the significant challenges and tradeoffs in terms of throughput, energy efficiency, and accuracy require a strongly interdisciplinary approach owing to the complex interplay of device technology, circuit engineering, and algorithm design.

Figure 1c shows a prototypical crosspoint-array-based accelerator for matrix-vector multiplication (MVM). Crosspoint arrays are compact memory ensembles where a group of devices is organized in a matrix fashion with top electrodes (TEs) connected column by column, and bottom electrodes (BEs) joined row by row, with a theoretically unrivaled bit cell size of $4F^2$ where F is the lithographic feature size in the process technology. The crosspoint array naturally provides an IMC hardware accelerator for analog, approximated MVM. Applying a voltage vector $\mathbf{v} = [V_1, V_2, \ldots, V_n]^T$ to columns of an array storing a conductance matrix $\mathbf{G}$, yields a current vector $\mathbf{i}$ at grounded rows:

$$\mathbf{i} = \mathbf{Gv}. \tag{1}$$

Computation is performed entirely in the analog domain, with Ohm and Kirchhoff laws providing physical equivalents for the multiply and accumulate operations used in digital computers. Thanks to the high parallelism of the crosspoint array structure, all required computations happen simultaneously, enabling $\mathcal{O}(1)$ MVM. Crosspoint MVM has been demonstrated for a broad range of problems, including image compression, sparse coding, and implementation of Artificial Neural Networks (ANNs), as well as hardware accelerators for algebraic operations [4].

Crosspoint memory arrays can be further enclosed in closed-loop feedback circuits to accelerate the solution of inverse problems, such as matrix inversion [5], eigenvector computation [6], and linear regression [7]. Figure 1d shows a closed-loop IMC (CL-IMC) circuit for linear system solution, where rows of a memory array storing conductance matrix $\mathbf{G}$ are connected to inverting input nodes of operational amplifiers (OAs), whereas columns are tied to the OA outputs. Thanks to the negative feedback, rows of the memory array are made virtual ground nodes of the circuit, allowing for the injection of a current vector $\mathbf{i}$. The same Kirchhoff current laws (KCL) of the open-loop MVM circuit hold, where the input/output role of currents $\mathbf{i}$ and output voltages $\mathbf{v}$ is reversed with respect to the MVM accelerator, yielding:

$$\mathbf{v} = -\mathbf{G}^{-1}\mathbf{i}. \tag{2}$$

CL-IMC retains $\mathcal{O}(1)$ complexity [8], with strong advantage over the typical $\mathcal{O}(n^\alpha)$ complexity of digital solvers, where n is the linear system matrix size and α is typically between 2 and 3 [9]. Consequently, CL-IMC is one of the most promising candidates for hardware acceleration of complex linear algebra computations. Among demonstrated applications are recommender systems such as Google's PageRank, with an estimated $100\times$ throughput improvement when compared with digital solvers [6], as well as machine learning (ML) techniques [7], by leveraging in-memory inversion of nonsquare matrices or *pseudoinversion*.

This chapter summarizes recent developments in closed-loop in-memory computing, ranging from compact modeling efforts towards a generalized framework, application investigation and circuit development, to experimental demonstrations in fully integrated testchips.

2 Modeling of Closed-Loop In-Memory Computing Circuits

A vast number of primitives have been conceived inside the in-memory framework, and either demonstrated experimentally or by numerical simulation [5–7, 10, 11]. To characterize proposed solutions, a wide array of ad-hoc models have been proposed with varying degree of generality [8, 12], although a unified model, capable of providing an eagle eye's view on the entire framework, was lacking. Similarly, ad-hoc simulators based on approximate models have been proposed [13], although with limited applicability to matrix-vector multiplication circuits only.

To model closed-loop IMC circuits, the general framework shown in Fig. 2a was introduced [14]. The framework is built around two model crosspoint arrays, namely the input array $\mathbf{Y}$ and feedback array $\mathbf{X}$. Loop closure is provided by an array of OAs whose inputs are connected to shared rows of $\mathbf{X}$ and $\mathbf{Y}$, and outputs are fed to columns of the feedback array $\mathbf{X}$. Row connection to the OAs can be performed either to the inverting or noninverting input. The opposite-polarity input of the operational amplifiers is instead grounded or connected to a reference voltage. Input signals can be provided either by current generators $\mathbf{i}$ connected to shared rows or by voltage generators $\mathbf{v_{in}}$ connected to columns of the input array $\mathbf{Y}$. The framework allows modeling of both MVM primitives (for which the operand matrix is encoded in the $\mathbf{Y}$ array, and the feedback array is set to the identity $\mathbf{I}$), IMVM primitives (for which the operand matrix is encoded in the $\mathbf{X}$ array, and the input array is set to the zero matrix $\mathbf{0}$), as well as more complex circuits through block-mapping [14].

The master equation for the generalized framework reads:

$$\frac{d\mathbf{v_{out}}}{dt} = \mathbf{T_0}^{-1}\big((\mathbf{SA_0}\hat{\mathbf{X}} - \mathbf{I})\mathbf{v_{out}}(t) + \mathbf{SA_0}\mathbf{U}^{-1}\mathbf{i}(t) + \mathbf{SA_0}\hat{\mathbf{Y}}\mathbf{v_{in}}(t)\big), \qquad (3)$$

where $\mathbf{v_{out}}$, $\mathbf{T_0}$, $\mathbf{S}$, $\mathbf{A_0}$, are the output voltage vector, and the OA time constant, sign, and gain (diagonal) matrices respectively. Matrix $\mathbf{U}$ represents the equivalent

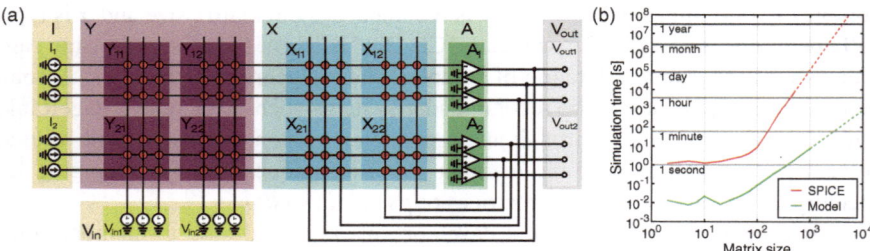

Fig. 2 Analytical modeling of inverse-matrix-vector multiplication circuits. **a** Generalized framework for the evaluation of crosspoint-based MVM and IMVM accelerators. **b** Wall-clock simulation time as a function of matrix size for SPICE solver (red) and analytical model in MatLab (green), demonstrating improvement in both absolute time and overall complexity. **a** and **b** Reproduced with permission from [14]. Copyright 2023 Author(s), licensed under a Creative Commons Attribution 4.0 License

conductance matrix seen at the shared rows [8], such that $\hat{\mathbf{X}}$, $\hat{\mathbf{Y}}$ are the conductive divider matrices describing transfers from the input and output voltage vectors to the OA input voltages, respectively.

Transformation of the master equation in the frequency domain yields the stability equation for the generalized framework, which, in the simplifying hypotheses of all OAs sharing the same gain α_0 and time constant τ_0, reduces to:

$$p = \frac{1}{\tau_0}(\alpha_0 \lambda_{S\hat{X}} - 1). \tag{4}$$

Following the Routh-Hurwitz stability criterion, for the circuit to exhibit a convergent response, the real part of the eigenvalues $\lambda_{S\hat{X}}$ of the signed divider matrix $\mathbf{S\hat{X}}$ should lie in the left-half plane (LHP). Since operand matrices do not generally meet this requisite in several applications, circuit topologies were designed [15] to lift the stability constraint and allow inversion of any matrix, irrespective of its eigenvalues.

Steady-state evaluation of the master equation allows assessment of circuit accuracy in the presence of different perturbation sources, including quantization of input sources and memory arrays, variability of memory arrays, and finite OA gain. Notably, the error introduced by each perturbation was found to be typically characterized by a perturbation-specific parameter, such as the number of bits for quantization and amplifier gain for virtual ground nonideality, and linearly dependent on the condition number κ_X of the feedback matrix [14], which can thus be recognized as the main sensitivity parameter for CL-IMC circuits.

Finally, the solution of the master equation in the time domain allows the derivation of a closed-form expression for the output voltage transient. Assuming outputs to be initially at rest, the output voltage reads:

$$\mathbf{v_{out}}(t) = -\left(\mathbf{I} - e^{\mathbf{T_0}^{-1}(\mathbf{SA_0}\hat{\mathbf{X}} - \mathbf{I})t}\right)(\mathbf{X} + \delta\mathbf{X}_\alpha)^{-1}(\mathbf{i} + \mathbf{Yv_{in}}). \tag{5}$$

Figure 2b compares the wall-clock simulation time as a function of matrix size for a SPICE solver and Eq. (5) evaluated in MatLab, highlighting a drastic reduction of the simulation time, as well as an improvement in the computational complexity from SPICE's $\mathcal{O}(n^4)$ down to $\mathcal{O}(n^\alpha)$, with $2 < \alpha < 3$ [14], allowing faster turnaround in the evaluation and study of IMC topologies.

3   Applications of Closed-Loop in-Memory Computing

Figure 3 shows a selection of the novel applications of CL-IMC that were demonstrated in recent years. Thanks to its $\mathcal{O}(1)$ computation, high energy efficiency, and high performance density, CL-IMC finds application in different domains of industrial interest, ranging from baseband processing in 6G communication systems [16], sensor fusion for autonomous driving [17], and data classification in machine learning [18].

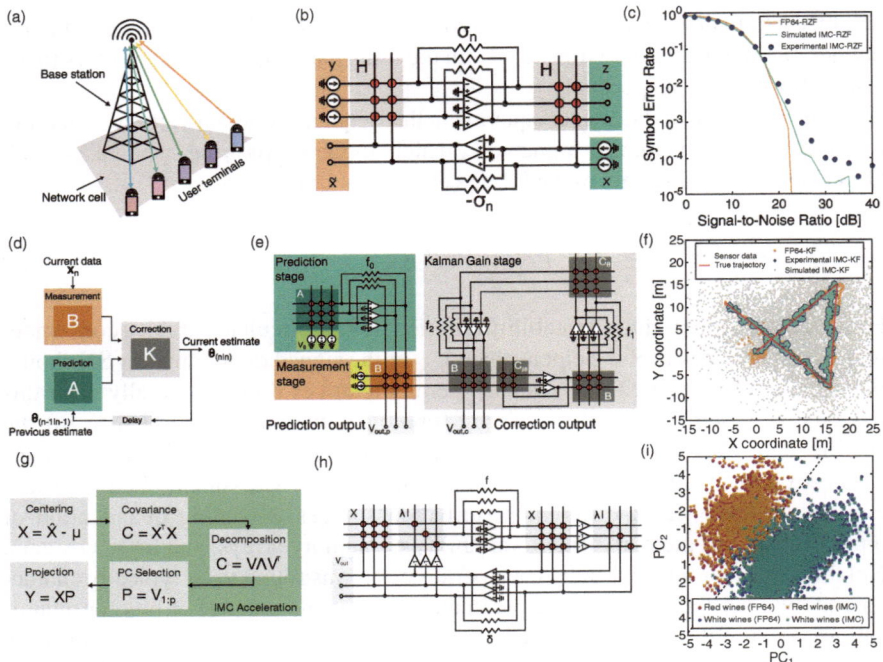

Fig. 3 Selected applications of CL-IMC. **a** Massive MIMO network cell, each comprising a single base station simultaneously serving many user terminals. **b** CL-IMC circuit for RZF decoding and precoding acceleration. **c** Symbol error rate as a function of channel signal-to-noise ratio for CL-IMC simulations (green), experiments (dark blue), and a fully-digital implementation (brown). **d** Kalman filter block diagram, consisting of prediction, measurement, and correction blocks each relying on a corresponding matrix. **e** CL-IMC circuit for Kalman filter. Each block is transposed to a corresponding in-memory circuit. **f** Object trajectories in the 2D-plane for a fully-digital FP64 Kalman filter (brown), simulations (green) and experiments (dark blue) of the CL-IMC topology, as compared to raw position measurement data (grey), and exact object trajectory (red). **g** Principal component analysis algorithm. **h** CL-IMC circuit for matrix eigendecomposition. **i** Wine dataset biplot obtained by CL-IMC (crosses) and by fully-digital FP64 (circles) PCA. **h** and **i** Reproduced with permission from [18]. Copyright 2024 Author(s), licensed under a Creative Commons Attribution 4.0 License

Figure 3a shows a typical network cell in massive multiple-input multiple-output (MIMO) systems, where a single *base station* (BS) equipped with an antenna array simultaneously serves many single-antenna *user terminals* (UTs). In a typical transaction, UTs first send their data **x** to the BS in the *uplink* phase. The propagation from the UTs to the BS is described by a *channel matrix* **H** and is affected by Gaussian white noise **w**, such that the overall received vector at the BS side is $\mathbf{y} = \mathbf{Hx} + \mathbf{w}$. The BS processes the received vector to compute an estimate $\tilde{\mathbf{x}}$ of the original message by regularized zero-forcing (RZF) according to:

$$\tilde{\mathbf{x}} = (\mathbf{H}^{\dagger}\mathbf{H} + \sigma_n^2\mathbf{I})^{-1}\mathbf{H}^{\dagger}\mathbf{y} \tag{6}$$

where σ_n^2 is the variance of $\mathbf{w}$. In the subsequent *downlink* phase, the BS replies to the users by transmitting a pre-coded version $\mathbf{z}$ of an intended message $\mathbf{x}$ accounting for channel distortion. Figure 3b shows a CL-IMC circuit for RZF decoding and precoding, where two memory arrays storing the channel matrix $\mathbf{H}$ are enclosed in a feedback loop through two transimpedance amplifier arrays, whose feedback conductances encode the regularization term $\sigma_n^2\mathbf{I}$. Input/output pairs $\{\mathbf{y}, \tilde{\mathbf{x}}\}$ and $\{\mathbf{x}, \mathbf{z}\}$ provide RZF decoding and precoding capability, respectively. Experimental validation was conducted on a fully-integrated testchip in 90 nm technology [17]. Figure 3c shows the symbol error rate (SER) as a function of the channel signal-to-noise ratio (SNR) for a 32-symbol Quadrature Amplitude Modulation (QAM) MIMO system for a CL-IMC system together with reference results for a fully-digital, 64-bit floating point (FP64) implementation, highlighting comparable accuracy with CL-IMC ultimately limited by the analog noise floor to SER $< 10^{-4}$.

Figure 3d shows the conceptual scheme of a Kalman filter (KF), an estimation algorithm for a set of unknown variables or *state* $\boldsymbol{\theta}$. KF operates iteratively, where a model $\mathbf{A}$ of the observed physical system provides a *prediction* of the state $\boldsymbol{\theta}$. Measurements $\mathbf{x}$, related to $\boldsymbol{\theta}$ by the measurement matrix $\mathbf{B}$ and its covariance $\mathbf{C}_{\beta\beta}$, are then used to refine the prediction by computing a *correction* term. The Kalman gain matrix defines the weight of the correction,

$$\mathbf{K} = \mathbf{C}_{\boldsymbol{\theta}}\mathbf{B}^T(\mathbf{B}\mathbf{C}_{\boldsymbol{\theta}}\mathbf{B}^T + \mathbf{C}_{\beta\beta})^{-1}, \tag{7}$$

such that the refined state estimate at time n $\boldsymbol{\theta}_{n|n}$, given the state at time $n-1$ $\boldsymbol{\theta}_{n-1|n-1}$, is given by:

$$\boldsymbol{\theta}_{n|n} = \mathbf{A}\boldsymbol{\theta}_{n-1|n-1} + \mathbf{K}(\mathbf{x}_n - \mathbf{B}\mathbf{A}\boldsymbol{\theta}_{n-1|n-1}). \tag{8}$$

Figure 3e shows a CL-IMC circuit for Kalman filter, where the prediction, measurement, and correction blocks are each implemented by a corresponding circuit section. Intermediate circuit outputs provide both the prediction and correction vectors, which can then be combined in the digital or analog domain. As a toy problem for circuit validation, two-dimensional path tracing was considered whereby an object moving along the xy plane is equipped with a noisy position sensor. The state vector comprises the object's position and velocity in the x and y directions. Figure 3f shows a trajectory composed of three direction changes together with position data sampled at 1 Hz frequency, and three different KF implementations, namely a fully-digital FP64, simulated CL-IMC and experimental CL-IMC [17]. CL-IMC results closely match FP64, allowing considerable improvement of the estimated trajectory with an ultimate position error $\sigma_{xy} \simeq \pm 1$ m.

Lastly, Fig. 3g shows the algorithm for Principal Component Analysis (PCA), a ML technique for dimensionality reduction. PCA computes the directions of maximum variance of a dataset, namely the *principal components* (PC), allowing projection in the PC space to highlight data correlation and perform classification and clustering. PC computation is performed by extracting eigenvectors and eigenvalues of the covariance matrix $\mathbf{C} = \mathbf{X}^T\mathbf{X}$, where $\mathbf{X}$ is a standardized version of the original dataset $\hat{\mathbf{X}}$. Dimensionality reduction is attained by retaining only PCs associated with

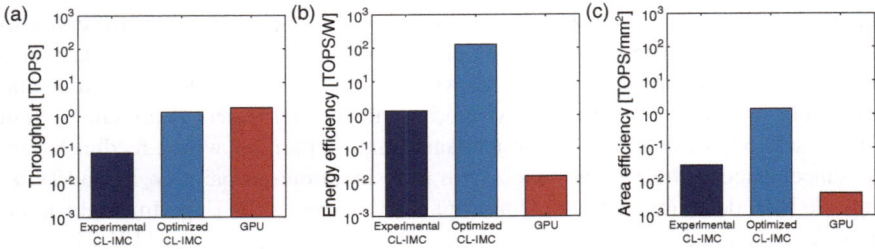

Fig. 4 Closed-loop in-memory computing benchmarks. **a** Throughput, **b** energy efficiency, and **c** area efficiency for a 90 nm CL-IMC testchip [14] (dark blue), a simulated CL-IMC primitive with optimized peripherals at 14 nm (light blue), and a reference graphic processing unit (red), benchmarked on a 64×64 matrix inversion task with $\mathcal{O}(n^3)$ complexity where n is the matrix size

the largest eigenvalues. Figure 3h shows a CL-IMC circuit for the computation of all eigenvectors and eigenvalues of a square matrix [18]. Four crosspoint arrays storing replicas of the operand matrix $\mathbf{X}$ and of the candidate eigenvalue $\lambda\mathbf{I}$ are feedback-connected using two transimpedance amplifiers. The circuit operates by positive feedback, exhibiting a self-sustained response if and only if arrays $\lambda\mathbf{I}$ match one of the eigenvalues of $\mathbf{X}$. By iteratively reprogramming $\lambda\mathbf{I}$ following a sweep-based procedure, all eigenvectors of $\mathbf{X}$ can be computed with minimal latency and energy consumption [18]. Figure 3i shows simulation results for computation of the first two PCs on the Wine dataset, for fully-digital FP64 and CL-IMC, yielding comparable classification accuracies of 98.32 and 98.08% respectively.

As a final figure of merit, Fig. 4 shows throughput, energy efficiency, and area efficiency benchmarks of CL-IMC for a 64×64 matrix inversion task, for an experimentally-measured 90 nm testchip [17], a simulated 14 nm testchip with optimized peripherals, and a reference graphic processing unit (GPU) [19]. Thanks to the IMC's $\mathcal{O}(1)$ advantage, closed-loop topologies provide comparable throughput with respect to state-of-the-art GPUs, while at the same time enabling $\times 10^4$ and $\times 10^2$ improvements in energy and area efficiency, supporting CL-IMC for fast and low-power machine learning on the edge.

4 Conclusions

In this chapter, recent developments in the study of closed-loop in-memory computing circuits have been presented. A generalized framework was developed to study IMC circuits, particularly providing fast evaluation of circuit dynamics by deriving a closed-form solution of the associated master equation. The framework was instrumental in the investigation of numerous applications, including baseband processing in 6G cellular networks, sensor fusion in autonomous driving, and data classification in machine learning. Novel circuit topologies were introduced and experimentally demonstrated on a dedicated testchip, achieving floating-point equivalent accuracy and throughput, with 2–4 orders of magnitude improvements in energy and area efficiency when compared to graphic processing units. The results obtained strengthen the position of closed-loop in-memory computing as a promising candidate for next-generation, energy-efficient machine learning accelerators in edge computing.

Acknowledgements The authors would like to thank E. Melacarne, E. Giannone, V. Pelleriti, G. Falcone, A. Pezzoli for their contributions to circuit characterization and testing, and G. Pedretti for fruitful discussions. Part of this work was supported by the European Research Council (ERC) (grant agreements no. 101054098 and ERC-2014-CoG-648635-RESCUE). Part of this work received funding from ECSEL Joint Undertaking (JU) under grant agreement no. 101007321.

References

1. von Neumann, J.: IEEE Ann. Hist. Comput. **15**(4), 27 (1993). https://doi.org/10.1109/85.238389
2. Ielmini, D., Wong, H.S.P.: Nat. Electron. **1**(6), 333 (2018). https://doi.org/10.1038/s41928-018-0092-2
3. Mannocci, P., Farronato, M., Lepri, N., Cattaneo, L., Glukhov, A., Sun, Z., Ielmini, D., Mach, A.P.L.: Learn. **1**(1), 010902 (2023). https://doi.org/10.1063/5.0136403
4. Zidan, M.A., Jeong, Y., Lee, J., Chen, B., Huang, S., Kushner, M.J., Lu, W.D.: Nat. Electron. **1**(7), 411 (2018). https://doi.org/10.1038/s41928-018-0100-6
5. Sun, Z., Pedretti, G., Ambrosi, E., Bricalli, A., Wang, W., Ielmini, D.: Proc. Natl. Acad. Sci. U.S.A. **116**(10), 4123 (2019). https://doi.org/10.1073/pnas.1815682116
6. Sun, Z., Ambrosi, E., Pedretti, G., Bricalli, A., Ielmini, D.: IEEE Trans. Electron. Dev. **67**(4), 1466 (2020). https://doi.org/10.1109/TED.2020.2966908
7. Sun, Z., Pedretti, G., Bricalli, A., Ielmini, D.: Sci. Adv. **6**(5), eaay2378 (2020) https://doi.org/10.1126/sciadv.aay2378
8. Sun, Z., Pedretti, G., Mannocci, P., Ambrosi, E., Bricalli, A., Ielmini, D.: IEEE Trans. Electron. Dev. **67**(7), 2945 (2020). https://doi.org/10.1109/TED.2020.2992435
9. Golub, G.H., Van Loan, C.F.: Matrix Computations. The Johns Hopkins University Press, Baltimore (2013). OCLC: ocn824733531
10. Li, C., Hu, M., Li, Y., Jiang, H., Ge, N., Montgomery, E., Zhang, J., Song, W., Dávila, N., Graves, C.E., Li, Z., Strachan, J.P., Lin, P., Wang, Z., Barnell, M., Wu, Q., Williams, R.S., Yang, J.J., Xia, Q.: Nat. Electron. **1**(1), 52 (2018). https://doi.org/10.1038/s41928-017-0002-z
11. Le Gallo, M., Sebastian, A., Mathis, R., Manica, M., Giefers, H., Tuma, T., Bekas, C., Curioni, A., Eleftheriou, E.: Nat. Electron. **1**(4), 246 (2018). https://doi.org/10.1038/s41928-018-0054-8
12. Sun, Z., Huang, R.: IEEE Trans. Circ. Syst. II: Express Briefs **68**(8), 2785 (2021). https://doi.org/10.1109/TCSII.2021.3068764
13. Chen, P.Y., Peng, X., Yu, S.: 2017 IEEE International Electron Devices Meeting (IEDM), pp. 6.1.1–6.1. IEEE, San Francisco, CA, USA (2017). https://doi.org/10.1109/IEDM.2017.8268337
14. Mannocci, P., Ielmini, D.: IEEE. J. Explor. Solid-State Comput. Dev. Circ. **9**(1), 47 (2023). https://doi.org/10.1109/JXCDC.2023.3265803
15. Mannocci, P., Pedretti, G., Giannone, E., Melacarne, E., Sun, Z., Ielmini, D.: IEEE Trans. Circ. Syst. I: Regul. Pap. **68**(12), 4889 (2021). https://doi.org/10.1109/TCSI.2021.3122278
16. Mannocci, P., Melacarne, E., Ielmini, D.: IEEE. J. Emerg. Sel. Top. Circ. Syst. **12**(4), 952 (2022). https://doi.org/10.1109/JETCAS.2022.3221284
17. Mannocci, P., Melacarne, E., Pezzoli, A., Pedretti, G., Villa, C., Sancandi, F., Spagnolini, U., Ielmini, D.: 2023 International Electron Devices Meeting (IEDM), pp. 1–4. IEEE, San Francisco, CA, USA (2023). https://doi.org/10.1109/IEDM45741.2023.10413724

18. Mannocci, P., Giannone, E., Ielmini, D.: IEEE Trans. Circ. Syst. II: Express Briefs **71**(4), 1839 (2024). https://doi.org/10.1109/TCSII.2023.3334958
19. NVIDIA RTX A2000 Embedded (2022). https://www.nvidia.com/en-us/design-visualization/resources/rtx-embedded/

Localized LO Phase Shifting for Phased Array Systems

Francesco Tesolin and **Salvatore Levantino**

Abstract Multiple-input-multiple-output is a promising technology to enable spatial multiplexing and improve throughput in wireless communication systems. The phase shifter needed in each element of the phased array to perform the electronic steering of the beam typically introduces a non-ideal transfer function in the signal path and consumes significant area and power, making it a major source of cost and dissipation. To address those issues, this chapter describes a novel technique referred to as localized LO phase-shifting, where the array of phase shifters in the receiver and in the transmitter is replaced by an array of synchronized PLLs, providing the local oscillation to each path, with fine and inherently linear phase regulation. This approach not only helps reduce power consumption and area occupation, in modern CMOS nodes, but also improves phase noise since the beam is formed by the combination of uncorrelated noise sources. To demonstrate the concept, a dual-element LO phase-shifting system, based on fractional-N digital PLLs in the 8.5-to-10.0-GHz range, is implemented in a standard 28-nm CMOS process. Each element occupies 0.23 mm^2 of area and dissipates 20 mW of power. An arbitrary phase shift between the LO outputs can be set over the full 360° range with a resolution of 0.7 millidegrees. The measured rms phase accuracy is 0.76°, and the peak-to-peak phase error is 2.1°, without requiring any linearity or gain calibration. Combining the outputs of the two elements, the measured integrated random jitter scales down from 58.5 to 44.6 fs rms.

1 Introduction

The continuous demand for larger data rates is pushing communication standards towards higher carrier frequencies (up to millimeter-wave), where larger bandwidths are available, but where significant propagation losses limit the maximum com-

F. Tesolin (✉) · S. Levantino
Politecnico di Milano, Piazza Leonardo da Vinci 32, 20133 Milan, Italy
e-mail: francesco.tesolin@polimi.it

S. Levantino
e-mail: salvatore.levantino@polimi.it

© The Author(s) 2025
S. Garatti (ed.), *Special Topics in Information Technology*,
PoliMI SpringerBriefs, https://doi.org/10.1007/978-3-031-80268-3_6

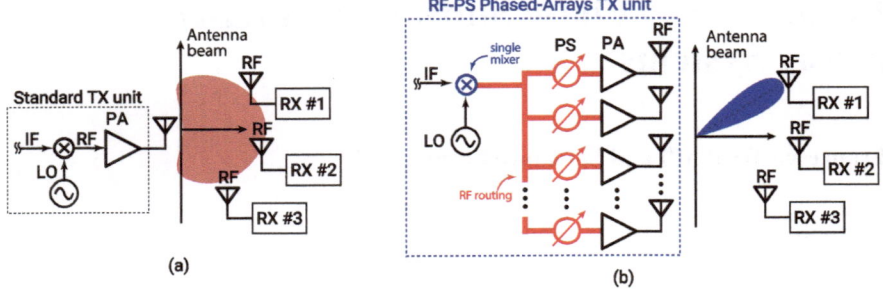

Fig. 1 Comparison between **a** a conventional radio unit with an omnidirectional antenna and **b** a phased array system with RF phase shifting, enabling beam steering

munication range. The radio implementation is therefore moving from the use of a single antenna with isotropic radiation patterns in Fig. 1a to the adoption of the multi-antenna approach in Fig. 1b, where the effective radiated power (ERP) in the direction of the desired receiver is increased, allowing for communication over larger distances. Additionally, the increased antenna directivity enables spatial multiplexing, minimizing the interference with nearby radios, and further increasing channel capacity. On the other hand, the larger antenna directivity requires accurate control of the beam direction, in order to track a specific user. By properly setting the relative phase shift between adjacent elements in the antenna array, it is possible to control the direction of the wavefront propagation relative to the antenna array plane. This is conventionally achieved by equipping each antenna path with a dedicated phase shifter (PS), as shown in Fig. 1b. The losses introduced by the PS and its phase inaccuracy directly affect the phased-array system performance, as well as its power consumption and footprint affect the overall system cost.

Various system-level approaches have been developed over the years to efficiently implement arrays of phase shifters, differing in the phase shifter's placement along the signal path. The most common topology, depicted in Fig. 1b, utilizes an RF-phase shifting architecture [1–9]. Here, the RF modulated carrier in each antenna element goes through a broadband RF PS. This design approach is compact as it requires only a single local oscillator (LO) and a mixer. However, routing N high-frequency signals (where N is the number of antenna elements) increases power consumption, and the high gain-versus-phase-shift sensitivity of the PS (because of the high-frequency operation) leads to a limited beam-steering accuracy. An alternative approach is to move the PS from the RF to the LO path, as depicted in Fig. 2a [10–15]. This approach necessitates one mixer per channel, but being the LO a signal with narrow bandwidth and constant-amplitude, the PS introduces no signal distortion and its gain-vs-phase-shift sensitivity plays no role. The main drawback of LO phase shifting is that distributing the high-frequency LO signal among the phased-array elements is challenging, as it has a large impact on the system power budget, practically limiting the maximum number of cores.

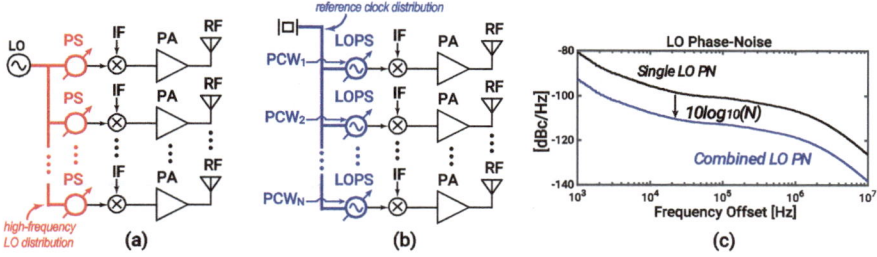

Fig. 2 **a** Traditional LO-phase-shifting, **b** proposed localized LO-phase-shifting architecture, and **c** phase-noise reduction with over-the-air combination of multiple LO elements

An alternative approach, namely the localized LO phase-shifting architecture, that greatly simplifies the LO distribution, is illustrated in Fig. 2b. Leveraging the low-power consumption, compact size, and excellent phase-noise performance of digital PLLs (DPLLs), the single LO in Fig. 2a is substituted with an array of DPLLs [16, 17]. The adoption of the direct phase modulation (DPM) technique in each DPLL allows fine phase-shift regulation [18]. Doing so, only the reference clock, originating from a crystal oscillator and typically operating in the 100-MHz frequency range, needs to be distributed to the elements of the phased-array, significantly relaxing the power consumption of the clock distribution network. Moreover, by embedding the phase-shifting capability in each LO element, no additional area or power consumption has to be accounted for the phase shifters. Furthermore, by employing one LO generator per each element of the transmitter array, the output beam resulting from the over-the-air combination of independent noise sources has a phase noise which improves with the number of element of the arrays, as shown in Fig. 2c. This means that a proportionally larger phase noise of each LO element can be tolerated and each element can be therefore scaled down in area occupation and power consumption.

2 Phase-Shifting Mechanism in a PLL Array

The implementation of the localized LO phase shifting technique is based on the use of a DPLL as LO of each element of the phased array. As DPLL, we will consider the type-II fractional-N digital PLL with bang-bang phase detector (BB-PD) depicted in the block diagram in Fig. 3a. Neglecting for a moment the signal $PCW[k]$, the DPLL generates an output frequency $f_{out} = FCW_0 \cdot f_{ref}$, where f_{ref} is the reference clock frequency and FCW_0 is a rational number, driving the modulus control of the multi-modulus divider (MMD) through a digital $\Delta\Sigma$ modulator. The $\Delta\Sigma$ modulator acts as a quantizer of FCW_0 and generates a sequence (with fewer bits) whose average value is proportional to FCW_0. The residual quantization error of the $\Delta\Sigma$ (i.e. the difference between the modulus control word and FCW_0) is canceled out by means of

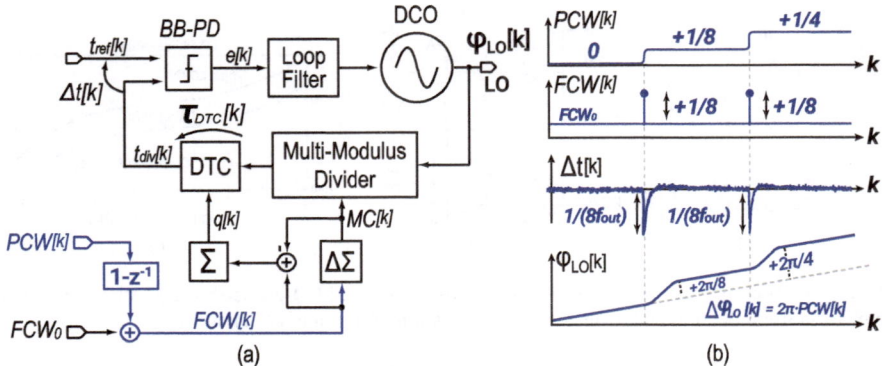

Fig. 3 **a** Digital phase-shifting of a fractional-N PLL with PCW input. **b** Transient waveforms after the application of the $PCW[k]$ sequence

a digital-to-time converter (DTC), after its amplitude is scaled with a gain calibrated in background by an LMS loop (not shown in figure) [19].

To accurately control the phase of the DPLL output at steady state, an additional phase-control-word $PCW[k]$ input is introduced. Since the phase of a periodic waveform is the derivative of the frequency, $PCW[k]$ goes through a first order difference before being added to FCW_0. In this way, the input $FCW[k]$ of the $\Delta\Sigma$ is given by

$$FCW[k] = FCW_0 + PCW[k] - PCW[k - 1]. \tag{1}$$

If $PCW[k]$ is increased step by step by a certain value, as in the signal diagram in Fig. 3b, $FCW[k]$, initially equal to FCW_0, undergoes a pulse (equal to the $PCW[k]$ step) for one reference clock cycle, before returning back to FCW_0. We want to derive the phase shift, $\Delta\varphi_{LO}$, asymptotically gained by the DPLL output signal after its transient. The timestamps of the zero crossings of the DTC output signal, $t_{div}[k]$, are given from the accumulation of the periods of the divider signal plus the DTC delay, and may be written as

$$t_{div}[k] = \sum_{i=0}^{k} MC[i] \cdot T_{dco}[i] + \tau_{DTC}[k], \tag{2}$$

where $MC[i]$ is the modulus control word at the ith time-stamp, $T_{dco}[i]$ the DCO time period (which will differ from the steady-state value of the period $T_{out} = 1/f_{out}$), while $\tau_{DTC}[k]$ is the DTC delay. Equation (2) can be rewritten highlighting the average value of the first term as follows

$$t_{div}[k] = T_{out} \cdot \sum_{i=0}^{k} MC[i] + \Delta t_{LO}[k] + q[k] \cdot T_{out}, \tag{3}$$

where $\Delta t_{LO}[k]$ is the time shift induced by the output frequency variations, and $q[k] \cdot T_{out}$ is the DTC delay sequence[1] cancelling the accumulated $\Delta\Sigma$ quantization noise $q[k]$. The latter is given by

$$q[k] = q[0] + \sum_{i=0}^{k}\Big(FCW[i] - MC[i]\Big) \tag{4}$$

$$= q[0] + k \cdot FCW_0 + PCW[k] - \sum_{i=0}^{k} MC[i]. \tag{5}$$

Plugging (5) in (3),

$$t_{div}[k] = \Delta t_{LO}[k] + q[0] \cdot T_{out} + k \cdot FCW_0 \cdot T_{out} + PCW[k] \cdot T_{out}. \tag{6}$$

Therefore, since we recognize $t_{ref}[k]$ in the third term of (6), and we can write the DPLL time error as

$$\Delta t[k] = t_{ref}[k] - t_{div}[k] = -PCW[k] \cdot T_{out} - \Delta t_{LO}[k] - q[0] \cdot T_{out}. \tag{7}$$

Since, in a type-II PLL, the time error $\Delta t[k]$ tends to zero at steady state, it follows that the asymptotic LO delay is given by

$$\Delta t_{LO}[k] = -T_{out} \cdot \Big(q[0] + PCW[k]\Big). \tag{8}$$

This time shift translates into an asymptotic LO output phase shift, after each $PCW[k]$ step, equal to

$$\Delta\varphi_{LO}[k] = 2\pi \cdot \Big(q[0] + PCW[k]\Big). \tag{9}$$

The result in (9) is theoretically relevant. It highlights two interesting properties: acting on $PCW[k]$, the LO phase shift at the end of the DPLL transient is *numerically stored* within the system. Thus, it is immune to disturbances and impairments of the analog circuits. Furthermore, the phase shift is *linearly dependent* on $PCW[k]$, suggesting that an accurate and controlled phase-shift can be induced in the LO, with extremely fine resolution. In fact, the phase-shifting resolution is only determined by the bit-length of the $\Delta\Sigma$ input. For instance, a 19-bit frequency control word (used in this work) enables a phase resolution of $360/2^{19} = 0.7 \cdot 10^{-3\circ}$.

---

[1] The T_{out} gain multiplying $q[k]$ is estimated by an LMS algorithm. As the LMS is much slower than the DPLL behavior, we can assume that the T_{out} gain is constant during DPLL transients.

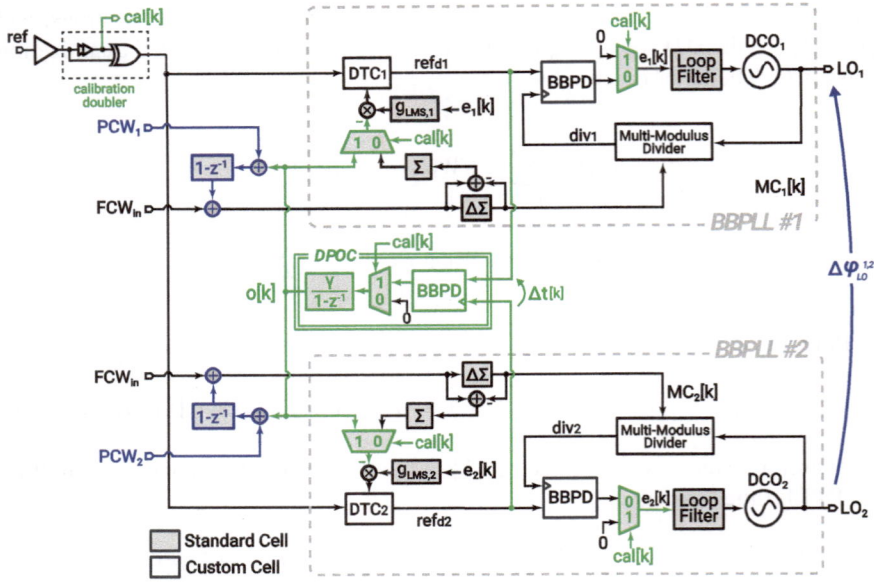

Fig. 4 Implemented dual-element LO-phase-shifting system: PCW input network (blue) and DPOC correction loop (green)

3   System Implementation and Measurement Results

Figure 4 shows the system block diagram of the two-element LO-phase-shifting system. Each DPLL has an output frequency spanning from 8.5 to 10 GHz, with 19-bit frequency resolution, while consuming 20-mW power. The DTC, along the reference path, is designed as in [20], with a 280-ps range and 13 bits (segmented in 7 coarse and 6 fine bits).

To force a certain phase shift between the two DPLLs through their phase-control words, PCW_1 and PCW_2, the two loops should exhibit the same initial value of the quantization error, see (9). However, a mismatch in the initial conditions exists, whenever the divider signal is used as clock of the digital network in each DPLL. This is because the number of initial clock cycles is affected by the specific DCO frequency transient of each DPLL. To solve this issue, both the $\Delta\Sigma$ modulators are clocked with the common reference clock [17].

Another aspect to consider in the phase-shift accuracy is the skew in the reference clock distribution. Even a skew of 3 ps in the reference clock would induce an output phase offset $>10°$ at 10 GHz. To make this contribution negligible, a digital offset correction technique (denoted as DPOC) is employed in the circuit [16, 17]. In practice, the phase of the two reference signals (after the DTCs) is compared with a second BB-PD, and a correction is fed back to both the PCW inputs. In this way, the phase offset of the two reference networks is cancelled out. The DPOC calibration

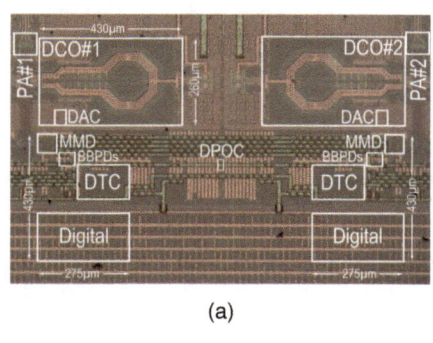
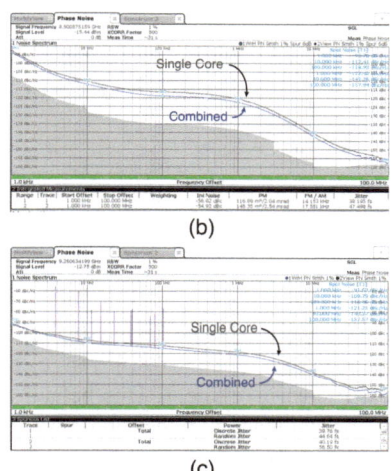

Fig. 5 Die micrograph **a** and measured phase noise spectra for **b** an integer-N channel and **c** a fractional-N near-integer channel, with single and combined element outputs

works in background, in alternate phase, according to the state of the $cal[k]$ signal. When $cal[k] = 1$, the DTCs of the two PLLs are disconnected and fed to the DPOC unit. When $cal[k] = 0$, instead, the normal PLL operation is restored. The timing of the $cal[k]$ signal is derived from a XOR-based frequency doubler circuit, which is also beneficial to lower jitter in normal operation mode.

The die photo of the implemented prototype in a 28-nm bulk CMOS technology is shown in Fig. 5a. The over-the-air combination of the LO outputs was emulated using a power combiner, and the phase noise measured at $\Delta PCW = 0$ is shown in Fig. 5b, in the case of an integer channel, and in Fig. 5c, in the case of a fractional near-integer channel. When switching from single- to dual-channel operation, the measured rms random jitter goes down from 47.5 to 38.2 fs (at the 8.50-GHz integer-N channel), and from 58.50 to 44.64 fs (at the fractional-N channel that is 3.8 kHz offset from the 9.25-GHz integer-N one).

The phase-shifting characteristic of the system, that is the output phase shift $\Delta\varphi_{LO}^{1,2}$ between the two DPLLs as a function of the differential phase-control word $\Delta PCW = PCW_1 - PCW_2$ is measured by means of a vector network analyzer (VNA). Two types of measurements were carried out. A "single channel" measurement obtained by varying only one of the two PCW inputs and assuming zero phase shift when $\Delta PCW = 0$, and a "dual channel" measurement by varying the PCW inputs in a pure differential mode. Sweeping ΔPCW over the $[-180, +180]°$ range, the phase shift of the single channel plotted in Fig. 6a spans in the $[-90, +90]°$ range, as expected. The phase error, computed as the difference between the measured phase-shift $\Delta\varphi_{LO}^{1,2}$ and the ideal programmed phase-shift (based on PCW) is plotted in Fig. 6b. It shows zero mean and random error with $0.21°$ peak-to-peak deviation

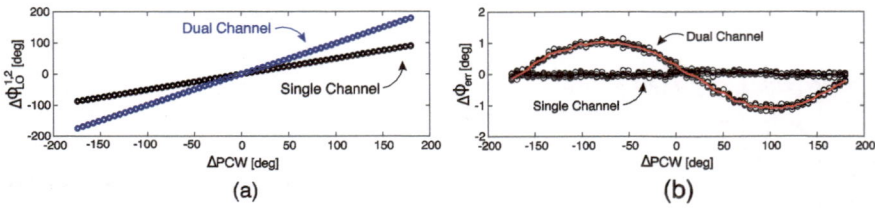

Fig. 6 Measured output phase characteristic **a** and phase error **b** versus the differential phase-control word ΔPCW, for the single- and the dual-channel case

and 0.044° rms value. In the dual-channel measurement, instead, although the phase spans in the $[-180, +180]°$ range correctly, a peculiar sinusoidal dependence of the phase error is evident. This behavior can be quantitatively justified by assuming a reciprocal leakage injection between the two channels, taking place downstream of the PLL loops, because of the coupling existing between the output power amplifiers, the wirebonds and the PCB lines [17]. Despite this undesired phenomenon, which can be reduced by improving decoupling, the peak-to-peak phase error is still only 2.1° with an rms error of 0.76°.

Table 1 compares the phase-shifting performance of the system with state-of-the-art LO phase-shifting systems. Note that the presented prototype reaches comparable

Table 1 Performance comparison of LO phase-shifting systems

	This work	K. Lee RFIC'22	J. Zhou CICC'22	H. J. Qian JSSC'21	Y. Wang JSSC'20	J. Pang JSSC'19	L. Wu JSSC'13
Phase-shifting architecture	**DPM-based Digital PLL**	PPF + Active Vector-Sum	PPF + Active Vector-Sum	Div. + Active Vector-Sum	PPF + Phase Sel. + LC tun.	PPF + Phase Sel. + LC tun.	Inj. locking oscillator
Technology node (nm)	**28**	40	40	40	65	65	65
Frequency range (GHz)	**8.5–10.0**	65.5–88.5	1.9–3.3	3–7	37–40	26.5–29.5	42.75–49.50
Phase range (°)	**360**	360	360	360	360	360	180
Resolution (°)/(bit)	**0.7 m/19b**	2.835/7b	0.35/10b	0.35/10b	0.04/13b	0.04/13b	22.5/4b
RMS/peak phase error (°)	**0.76/2.1**	0.29/N.A.	0.4/1.0	1.6/2.3	0.08/0.5	0.3/5.1	0.93/1.5
Single element power (mW)	**20.0[a]**	101.9	1330[b]	15.4	60	10.3	21.25
Single element area (mm²)	**0.23[a]**	0.17	2.145[b]	0.05	3	0.25	0.7
Linearity/gain calibration	Not needed	Off-Chip LUT	On-Chip analog	Off-Chip DPD	On-Chip analog	Off-Chip LUT	On/Off-Chip analog

[a]Includes LO generation
[b]Includes PA

accuracy in terms of rms/peak-error *without adopting any foreground or background gain/linearity calibration* and provides the highest phase-shifting resolution (19 bits) while keeping compact area and low power consumption.

4 Conclusion

This chapter has presented a novel promising technique to implement phased-array systems which reduces area and power consumption, namely the localized LO-phase-shifting technique. The technique replaces the conventional array of phase shifters along the signal paths with an array of synchronized digital PLLs with direct phase modulation, acting as LO generators. Analysis has shown that the PLLs can be synchronized in phase and their individual phase can be digitally controlled with high resolution and ideal linearity. A dual-element localized LO phase-shifting system has been implemented and fabricated in 28-nm CMOS. Phase-shift between the LO elements with 19-bit resolution and 0.76° rms phase error has been demonstrated with no extra area/power consumption and with no gain/linearity calibration. Compared with other state-of-the-art LO-phase-shifting schemes with similar phase accuracy, the implemented system exhibits the lowest area occupation and power consumption, making the proposed architecture appropriate for massive MIMO applications.

References

1. Yi, Y., Zhao, D., Zhang, J., Gu, P., Chai, Y., Liu, H., You, X.: A 24-29.5-GHz highly linear phased-array transceiver front-end in 65-nm CMOS supporting 800-MHz 64-QAM and 400-MHz 256-QAM for 5G new radio. IEEE J. Solid-State Circ. **57**(9), 2702–2718 (2022)
2. Alhamed, A., Gültepe, G., Rebeiz, G.M.: A multi-band 16-52-GHz transmit phased array employing 4 × 1 beamforming IC with 14-15.4-dBm Psat for 5G NR FR2 operation. IEEE J. Solid-State Circ. **57**(5), 1280–1290 (2022)
3. Zhu, W., Wang, J., Zhang, X., Lv, W., Liao, B., Zhu, Y., Wang, Y.: A 24-28-GHz four-element phased-array transceiver front end with 21.1%/16.6% transmitter peak/OP1dB PAE and subdegree phase resolution supporting 2.4 Gb/s in 256-QAM for 5-G communications. IEEE Trans. Microwave Theory Tech. **69**(6), 2854–2869 (2021)
4. Li, M., Li, N., Gao, H., Wang, S., Zhang, Z., Chen, P., Wei, N., Gu, Q.J., Song, C., Xu, Z.: 14.7 an adaptive analog temperature-healing low-power 17.7-to-19.2GHz RX front-end with ±0.005dB/°C gain variation, <1.6dB NF variation, and <2.2dB IP1dB variation across −15 to 85°C for phased-array receiver. In: 2021 IEEE International Solid- State Circuits Conference (ISSCC), vol. 64, pp. 230–232 (2021)
5. Khalil, A., Eshrah, I., Elsherief, A., Mehana, A., Abdalla, M., Mobarak, M., Kilpatrick, J., Hall, B., Ashry, A., Fahmy, H., Salim, S., Kernan, R., Herdeg, B., Sapia, G., El-Nozahi, M., Weheiba, M., D'Amato, M., Bautista, C., Chatzopoulos, K., Ghoniem, A., Mosa, Y., Roll, D., Ok, K.: 2.1 mm-Wave 5G radios: baseband to waves. In: 2021 IEEE International Solid-State Circuits Conference (ISSCC), vol. 64, pp. 38–40 (2021)

6. Pang, J., Li, Z., Kubozoe, R., Luo, X., Wu, R., Wang, Y., You, D., Fadila, A.A., Saengchan, R., Nakamura, T., Alvin, J., Matsumoto, D., Liu, B., Narayanan, A.T., Qiu, J., Liu, H., Sun, Z., Huang, H., Tokgoz, K.K., Motoi, K., Oshima, N., Hori, S., Kunihiro, K., Kaneko, T., Shirane, A., Okada, K.: A 28-GHz CMOS phased-array beamformer utilizing neutralized bi-directional technique supporting dual-polarized MIMO for 5G NR. IEEE J. Solid-State Circ. **55**(9), 2371–2386 (2020)

7. Kibaroglu, K., Sayginer, M., Rebeiz, G.M.: A low-cost scalable 32-element 28-GHz phased array transceiver for 5G communication links based on a 2 × 2 beamformer flip-chip unit cell. IEEE J. Solid-State Circ. **53**(5), 1260–1274 (2018)

8. Kim, H.-T., Park, B.-S., Oh, S.-M., Song, S.-S., Kim, J.-M., Kim, S.-H., Moon, T.-S., Kim, S.-Y., Chang, J.-Y., Kim, S.-W., Kang, W.-S., Jung, S.-Y., Tak, G.-Y., Du, J.-K., Suh, Y.-S., Ho, Y.-C.: A 28GHz CMOS direct conversion transceiver with packaged antenna arrays for 5G cellular system. In: 2017 IEEE Radio Frequency Integrated Circuits Symposium (RFIC), pp. 69–72 (2017)

9. Sadhu, B., Tousi, Y., Hallin, J., Sahl, S., Reynolds, S.K., Renström, Ö., Sjögren, K., Haapalahti, O., Mazor, N., Bokinge, B., Weibull, G., Bengtsson, H., Carlinger, A., Westesson, E., Thillberg, J.-E., Rexberg, L., Yeck, M., Xiaoxiong, G., Ferriss, M., Liu, D., Friedman, D., Valdes-Garcia, A.: A 28-GHz 32-element TRX phased-array IC with concurrent dual-polarized operation and orthogonal phase and gain control for 5G communications. IEEE J. Solid-State Circ. **52**(12), 3373–3391 (2017)

10. Lee, K., Choi, C.-G., Kim, K., Lee, S., Choi, S.-U., Lee, J., Koo, B., Song, H.-J.: Highly accurate frequency quadrupler based lo phase shifter achieving 0.29° RMS phase error for wideband e-band beamforming receiver. In: 2022 IEEE Radio Frequency Integrated Circuits Symposium (RFIC), pp. 323–326 (2022)

11. Zhou, J., Qian, H.J., Yang, B., Shu, Y., Luo, X.: A phase-modulation phase-shifting phased-array transmitter with 10-bit fast-locking phase self-calibration and 0/2.5/6/12dB power back-offs efficiency enhancement. In: 2022 IEEE Custom Integrated Circuits Conference (CICC), pp. 01–02 (2022)

12. Qian, H.J., Zhou, J., Yang, B., Luo, X.: A 4-element digital modulated polar phased-array transmitter with phase modulation phase-shifting. IEEE J. Solid-State Circ. **56**(11), 3331–3347 (2021)

13. Wang, Y., Wu, R., Pang, J., You, D., Fadila, A.A., Saengchan, R., Fu, X., Matsumoto, D., Nakamura, T., Kubozoe, R., Kawabuchi, M., Liu, B., Zhang, H., Qiu, J., Liu, H., Oshima, N., Motoi, K., Hori, S., Kunihiro, K., Kaneko, T., Shirane, A., Okada, K.: A 39-GHz 64-element phased-array transceiver with built-in phase and amplitude calibrations for large-array 5G NR in 65-nm CMOS. IEEE J. Solid-State Circ. **55**(5), 1249–1269 (2020)

14. Pang, J., Wu, R., Wang, Y., Dome, M., Kato, H., Huang, H., Tharayil Narayanan, A., Liu, H., Liu, B., Nakamura, T., Fujimura, T., Kawabuchi, M., Kubozoe, R., Miura, T., Matsumoto, D., Li, Z., Oshima, N., Motoi, K., Hori, S., Kunihiro, K., Kaneko, T., Shirane, A., Okada, K.: A 28-GHz CMOS phased-array transceiver based on LO phase-shifting architecture with gain invariant phase tuning for 5G new radio. IEEE J. Solid-State Circ. **54**(5), 1228–1242 (2019)

15. Wu, L., Li, A., Luong, H.C.: A 4-Path 42.8-to-49.5 GHz LO generation with automatic phase tuning for 60 GHz phased-array receivers. IEEE J. Solid-State Circ. **48**(10), 2309–2322 (2013)

16. Santiccioli, A., Mercandelli, M., Dartizio, S.M., Tesolin, F., Karman, S., Shehata, A., Bertulessi, L., Buccoleri, F., Avallone, L., Parisi, A., Cherniak, D., Lacaita, A.L., Kennedy, M.P., Samori, C., Levantino, S.: 32.8 A 98.4fs-Jitter 12.9-to-15.1GHz PLL-based LO phase-shifting system with digital background phase-offset correction for integrated phased arrays. In: 2021 IEEE International Solid- State Circuits Conference (ISSCC), vol. 64, pp. 456–458 (2021)

17. Tesolin, F., Dartizio, S.M., Buccoleri, F., Santiccioli, A., Bertulessi, L., Samori, C., Lacaita A.L., Levantino, S.: A novel LO phase-shifting system based on digital bang-bang PLLs with background phase-offset correction for integrated phased arrays. IEEE J. Solid-State Cir. **58**(9), 2466–2477 (2023)

8. Marzin, G., Levantino, S., Samori, C., Lacaita, A.L.: A 20 Mb/s phase modulator based on 3.6 GHz digital PLL with −36 dB EVM at 5 mW power. IEEE J. Solid-State Circ. **47**(1 2974–2988 (2012)

19. Tasca, D., Zanuso, M., Marzin, G., Levantino, S., Samori, C., Lacaita, A.L.: A 2.9-4.0-GHz fractional-N digital PLL with bang-bang phase detector and 560-fs$_{rms}$ integrated Jitter at 4.5-mW power. IEEE J. Solid-State Circ. **46**(12), 2745–2758 (2011)
20. Santiccioli, A., Mercandelli, M., Bertulessi, L., Parisi, A., Cherniak, D., Lacaita, A.L., Samori, C., Levantino, S.: A 66-fs-rms Jitter 12.8-to-15.2-GHz fractional-N bang-bang PLL with digital frequency-error recovery for fast locking. IEEE J. Solid-State Circ. **55**(12), 3349–3361 (2020)

Systems and Control

Data-Based Control Design for Recurrent Neural Network Models with Stability Guarantees

William D'Amico⬤

Abstract This Brief aims to discuss the application of data-driven methods to control systems described by recurrent neural network models, which are known to be universal approximators of dynamical systems. The unified hybrid approach outlined here fills a significant gap in the current literature by combining the strengths of both direct and indirect methodologies so as to ensure in a purely data-driven fashion, on the one hand, desired performances and, on the other hand, closed-loop stability.

1 Introduction

In control systems engineering, data-based techniques are gaining increasing attention due to the ever-growing availability of data in many large scale applications and processes (e.g., chemical industry, metallurgy, machinery, electronics, electricity, transportation, and logistics), and since they allow one to design performing controllers with moderate time and computational effort. They are mainly divided in (i) direct methods, where a controller is directly identified from data through optimization, and (ii) indirect methods, where a model of the system is firstly identified from data, based on which the controller is then designed.

Among direct methods, virtual reference feedback tuning (VRFT) has gained wide popularity due to its simplicity and satisfactory performances [1]. One of the major issues in VRFT, as well as in many direct data-driven methods, concerns the possibility of providing stability guarantees for the closed-loop system. In fact, based on an available batch of data, the VRFT method only allows one to design a controller such that the control system is as close as possible to a given reference closed-loop model. However, especially if the controller identification results are poor and the obtained regulator is far from the ideal one, the resulting feedback system may display bad performances and even instability. In the literature, only "a-posteriori" validation

W. D'Amico (✉)
Dipartimento di Elettronica, Informazione e Bioingegneria, Politecnico di Milano, Via Ponzio 34/5, 20133 Milano, Italy
e-mail: william.damico@polimi.it

© The Author(s) 2025
S. Garatti (ed.), *Special Topics in Information Technology*,
PoliMI SpringerBriefs, https://doi.org/10.1007/978-3-031-80268-3_7

69

tests, aiming at verifying the closed-loop stability, have been proposed in the linear setting [2, 3].

The ever-growing amount of informative data paved the way also to the use of neural networks (NNs) to learn features from data. For this reason, NNs have gained great interest in many engineering fields, among which control applications. They can be divided in feedforward neural networks (FFNNs), to reproduce static systems, and recurrent neural networks (RNNs), characterized by the presence of internal loops representing state variables allowing them to replicate dynamical systems. Notably, RNNs have been used for designing controllers applicable to dynamical systems in both indirect and direct frameworks. For example, in [4], a model predictive control approach based on an identified long short-term memory (LSTM) network is exploited to control the cooling system of a large business center. Moreover, in [5], controllers based on echo state networks (ESNs) and LSTMs are directly learned via VRFT to control an experimental electronic throttle body.

Despite the popularity and potentialities of RNNs, their theoretical properties have been marginally analysed in the literature. As nonlinear dynamical systems, it is in fact fundamental to characterize conditions that guarantee the stability of their motions, especially when RNNs are part of control systems. In this context, incremental input-to-state stability (δISS) [6] plays a crucial role. This property entails that the dynamics of a δISS RNN is asymptotically independent of its initialization and implies other common stability properties such as global asymptotic stability (GAS) of the equilibria. Interestingly, the δISS property can be directly enforced in the RNN training phase [7, 8]. However, the latter works focus on open-loop RNNs and they do not address the design of stabilizing RNN-based feedback controllers.

Regarding control systems, regulation strategies based on model predictive control are presented for some RNN architectures in [9, 10], ensuring closed-loop stability if the RNN-based model of the controlled system enjoys the δISS property. Finally, less conservative local asymptotic stability guarantees for an equilibrium of the feedback system are investigated in [11, 12], where event-triggered FFNN controllers are employed and an estimate of the domain of attraction is also provided.

1.1 Contributions

The work herein contained summarizes some contributions given in [13], where a novel unified framework is proposed for the application of data-based methods to control RNN-based systems while providing closed-loop stability guarantees. The two main contributions here outlined are the following.

A δISS condition for a class of RNNs—Firstly, in [14], a novel condition for δISS, proved to be less conservative than other conditions available in the literature, is devised for a general class of RNNs including, e.g., ESNs [15], shallow neural nonlinear autoregressive exogenous (NNARX) models [7], and the "recurrent nets" considered in [16, Eq. (2)].

Control design of RNN-based closed-loop systems—Then, in [17], by considering an identified RNN model for the system, the previous δISS condition is rearranged in the form of linear matrix inequalities (LMIs) to tune ESN controllers and compensators guaranteeing the δISS of the closed-loop system, where an explicit integral action is also embedded. The integration of this condition into unifying LMI-based optimization problems is also explored, where a performance metric is enforced via an LMI reformulation of VRFT, initially proposed in [18, 19] for the linear setting.

The reader is referred to [13, Sect. 1.5] for the main notation adopted throughout this chapter, as well as for some preliminary definitions.

2 An Incremental Input-to-State Stability Condition for a Class of RNNs

As previously discussed, the δISS property is a relevant property for RNN models since it implies a number of useful features and stability properties (e.g., GAS of the equilibria). In particular, we focus on the class of RNNs introduced in the following section, which includes, e.g., ESNs and shallow NNARXs.

2.1 RNN System Model

We consider the following class of shallow recurrent neural networks defined by

$$x(k+1) = f(Ax(k) + Bu(k)), \tag{1a}$$

$$y(k) = Cx(k) + Du(k), \tag{1b}$$

where $u \in \mathbb{R}^m$ is the exogenous variable, $y \in \mathbb{R}^l$ is the output vector, $x \in \mathbb{R}^n$ is the state vector, $f(\cdot) = \begin{bmatrix} f_1(\cdot) & \dots & f_n(\cdot) \end{bmatrix}^\top \in \mathbb{R}^n$ is a vector of scalar functions applied element-wise, $A \in \mathbb{R}^{n \times n}$, $B \in \mathbb{R}^{n \times m}$, $C \in \mathbb{R}^{l \times n}$, and $D \in \mathbb{R}^{l \times m}$. The function $f(\cdot)$ must fulfill the following assumption.

Assumption 1 The functions $f_i(\cdot)$, $i = 1, \dots, n$, are nonlinear globally Lipschitz continuous functions with Lipschitz constant L_{pi} or the identity function $\mathrm{id}(\cdot)$.

Since the function $f(\cdot)$ in class (1) may contain also some identity functions, let us introduce the set

$$\mathcal{W} := \{i \in \{1, \dots, n\} \mid f_i(\cdot) \neq \mathrm{id}(\cdot)\}. \tag{2}$$

Note that the possible presence of some identity functions can be exploited later in the δISS condition to provide more degrees of freedom and to reduce conservatism.

2.2   A Novel Sufficient Condition for Incremental
Input-to-State Stability of Shallow RNNs in Class (1)

Let us consider a generic system (1) fulfilling Assumption 1. The following theorem provides a sufficient condition which guarantees the δISS for nonlinear systems lying in the class (1a). We first define a diagonal matrix

$$W := \operatorname{diag}(L_{p1}, \dots, L_{pn}) \in \mathbb{D}^n \,,$$

where $L_{pi} = 1$ for all $i \notin \mathcal{W}$. We introduce the following matrices:

$$\widetilde{A} := W A \,, \qquad \widetilde{B} := W B \,.$$

Theorem 1 *Let Assumption 1 hold. System (1a) is δISS if $\exists P \in \mathbb{S}_{>0}^n$ such that $p_{ij} = p_{ji} = 0 \; \forall i \in \mathcal{W}, \; \forall j \in \{1, \dots, n\}$ with $j \neq i$, and*

$$\widetilde{A}^\top P \widetilde{A} - P \prec 0 \,. \tag{3}$$

The reader is referred to [13, Theorem 7.2.1] for the proof of this theorem. Theorem 1 ensures that system (1a) is δISS if $\widetilde{A}$ is Schur stable and if there exists a structured matrix P fulfilling the Lyapunov inequality (3). Specifically, P must have zero elements along all the rows and columns (except for the diagonal element) corresponding to the rows of system (1a) whose activation function is nonlinear.

3   Control Design of RNN-Based Closed-Loop Systems

In this section, the previously introduced condition for δISS of RNN systems (1) is exploited to tune controllers guaranteeing δISS properties for the RNN-based closed-loop system depicted in Fig. 1. We consider a nonlinear discrete-time system $\mathcal{S}$, possibly obtained via an identification procedure, defined by

$$x_s(k + 1) = f_s(A_s x_s(k) + B_s u_s(k)) \,, \tag{4a}$$
$$y_s(k) = C_s x_s(k) \,, \tag{4b}$$

where $u_s \in \mathbb{R}^{m_s}$ is the exogenous variable, $y_s \in \mathbb{R}^{l_s}$ is the output vector, $x_s \in \mathbb{R}^{n_s}$ is the state vector, $f_s(\cdot) = \left[f_{s,1}(\cdot) \; \cdots \; f_{s,n_s}(\cdot) \right]^\top \in \mathbb{R}^{n_s}$ is a vector of functions fulfilling Assumption 1 applied element-wise, $A_s \in \mathbb{R}^{n_s \times n_s}$, $B_s \in \mathbb{R}^{n_s \times m_s}$, and $C_s \in \mathbb{R}^{l_s \times n_s}$. Let $W_s \in \mathbb{D}^{n_s}$ contain the Lipschitz constants on the main diagonal.

To make the problem solvable with a lightweight LMI optimization one, a feedback controller C and a feedforward compensator $\mathcal{H}$ are selected in the class of ESNs, with equations

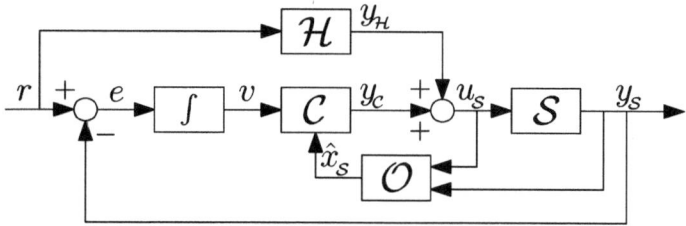

Fig. 1 Control scheme: C is a feedback ESN controller, "$\int$" is a discrete-time integrator, $\mathcal{H}$ is a feedforward ESN compensator, and O is a state observer

$$x_\ell(k+1) = \zeta_\ell(W_{x_\ell} x_\ell(k) + W_{u_\ell} u_\ell(k) + W_{y_\ell} y_\ell(k)), \tag{5a}$$

$$y_\ell(k) = V_{\text{out}_{1,\ell}} x_\ell(k) + V_{\text{out}_{2,\ell}} u_\ell(k), \tag{5b}$$

where subscript $\ell \in \{\mathcal{H}, C\}$ is used to refer to the compensator $\mathcal{H}$ and to the feedback controller C, respectively. As discussed in [9], the design of $\mathcal{H}$ and C consists of the tuning of the matrices $V_{\text{out}_{1,\ell}}$ and $V_{\text{out}_{2,\ell}}$, for $\ell \in \{\mathcal{H}, C\}$, whereas W_{x_ℓ}, W_{u_ℓ}, and W_{y_ℓ} are fixed and ramdomly generated matrices. In Eq. (5), $u_C(k) = \left[v(k)^\top \; \hat{x}_s(k)^\top\right]^\top$, $u_{\mathcal{H}}(k) = r(k)$, the functions $\zeta_{\ell,i}(\cdot)$ of $\zeta_\ell(\cdot)$ are globally Lipschitz continuous $\forall i = 1, \ldots, n_\ell$, $W_{u_{\mathcal{H}}} \in \mathbb{R}^{n_{\mathcal{H}} \times l_s}$, $V_{\text{out}_{2,\mathcal{H}}} \in \mathbb{R}^{m_s \times l_s}$, $W_{x_\ell} \in \mathbb{R}^{n_\ell \times n_\ell}$, $W_{y_\ell} \in \mathbb{R}^{n_\ell \times m_s}$, and $V_{\text{out}_{1,\ell}} \in \mathbb{R}^{m_s \times n_\ell}$. Let us define $W_{u_C} := \left[W_{u_C}^v \; W_{u_C}^x\right]$ and $V_{\text{out}_{2,C}} := \left[V_{\text{out}_{2,C}}^v \; V_{\text{out}_{2,C}}^x\right]$, where $W_{u_C}^v \in \mathbb{R}^{n_C \times l_s}$, $W_{u_C}^x \in \mathbb{R}^{n_C \times n_s}$, $V_{\text{out}_{2,C}}^v \in \mathbb{R}^{m_s \times l_s}$, and $V_{\text{out}_{2,C}}^x \in \mathbb{R}^{m_s \times n_s}$. The discrete-time integrator block "$\int$" has equation

$$\eta(k+1) = \eta(k) + e(k), \tag{6a}$$

$$v(k) = \eta(k) + e(k), \tag{6b}$$

where $e(k) = r(k) - y_s(k)$. Finally, the block O denotes a state observer defined by

$$\hat{x}_s(k+1) = f_s(A_s \hat{x}_s(k) + B_s u_s(k) + \hat{L}(y_s(k) - \hat{y}_s(k))), \tag{7a}$$

$$\hat{y}_s(k) = C_s \hat{x}_s(k), \tag{7b}$$

where $\hat{y}_s \in \mathbb{R}^{l_s}$ is the predicted output vector, $\hat{x}_s \in \mathbb{R}^{n_s}$ is the predicted state vector, and the observer gain $\hat{L} \in \mathbb{R}^{n_s \times l_s}$ is a design parameter.

The objective is the design of $\mathcal{H}$, C, and O so as to (i) provide global asymptotic stability guarantees for the equilibria of the control system, (ii) achieve asymptotic tracking of constant reference signals r, and (iii) make the control system as similar as possible to an "invertible" reference model $\mathcal{M}$.

3.1 Observer Design

Firstly, the result in Theorem 1 can be exploited to design a convergent observer O as in Eq. (7) enjoying the δISS property. The observer gain $\hat{L}$ can be obtained by solving the optimization problem specified in [13, Lemma 10.2.1].

3.2 Stability Guarantees

Then, the δISS of the control scheme in Fig. 1 is ensured by conferring, with two separated LMIs, δISS to both $\mathcal{H}$ and the closed-loop system obtained by discarding $\mathcal{H}$. This follows from the properties of the cascade of δISS systems. In this section, we recall only the LMI ensuring the δISS of the closed-loop system without $\mathcal{H}$, whereas the reader is referred to [13, Lemma 10.2.3] for the LMI related to the δISS of $\mathcal{H}$.

We consider the closed-loop system in Fig. 1, where $\mathcal{H}$ is discarded and $y_{\mathcal{H}}$ acts as an independent exogenous variable. Note also that $\hat{x}_s(k) = x_s(k) - \hat{e}(k)$, where the observer error $\hat{e}$ acts as an asymptotically vanishing perturbation, provided that the observer is convergent. By jointly considering Eqs. (4)–(6), we can write the closed-loop system as in Eq. (1a), where Assumption 1 is satisfied, and

$$x(k) = \begin{bmatrix} x_c(k) \\ \eta(k) \\ x_s(k) \end{bmatrix}, \quad u(k) = \begin{bmatrix} r(k) \\ y_{\mathcal{H}}(k) \\ \hat{e}(k) \end{bmatrix}, \quad f(\cdot) = \begin{bmatrix} \zeta_c(\cdot) \\ \mathrm{id}_{l_S}(\cdot) \\ f_s(\cdot) \end{bmatrix}.$$

In particular, $A = F_c + G_c J_c$, where

$$F_c := \begin{bmatrix} W_{x_c} & W^v_{u_c} & W^x_{u_c} - W^v_{u_c} C_S \\ 0 & I & -C_S \\ 0 & 0 & A_s \end{bmatrix}, \quad G_c := \begin{bmatrix} W_{y_c} \\ 0 \\ B_s \end{bmatrix},$$

$J_c := \begin{bmatrix} V_{\mathrm{out}_{1,c}} & V^v_{\mathrm{out}_{2,c}} & V^x_{\mathrm{out}_{2,c}} - V^v_{\mathrm{out}_{2,c}} C_S \end{bmatrix} = \begin{bmatrix} V_{\mathrm{out}_{1,c}} & V_{\mathrm{out}_{2,c}} \end{bmatrix} E$, and

$$E := \begin{bmatrix} I & 0 & 0 \\ 0 & I & -C_S \\ 0 & 0 & I \end{bmatrix} \tag{8}$$

is an invertible matrix. Hence, Theorem 1 can be applied to enforce the δISS of the closed loop, according to the following lemma proved in [13, Lemma 10.2.2].

Lemma 1 Let $\mathcal{W} := \{i \in \{1, \dots, \tilde{n}\} \mid f_i(\cdot) \neq \mathrm{id}(\cdot)\}$ and $\tilde{n} := n_s + n_c + l_s$. If $\exists\, P_c = P_c^\top \in \mathbb{R}^{\tilde{n} \times \tilde{n}}$ such that $(p_c)_{ij} = (p_c)_{ji} = 0\ \forall i \in \mathcal{W},\ \forall j \in \{1, \dots, \tilde{n}\}$ with $j \neq i$, and $\exists\, L_c \in \mathbb{R}^{m_s \times \tilde{n}}$, $Q_c \in \mathbb{R}^{\tilde{n} \times \tilde{n}}$ such that

$$\begin{bmatrix} P_c & \widetilde{F}_c Q_c + \widetilde{G}_c L_c \\ (\widetilde{F}_c Q_c + \widetilde{G}_c L_c)^\top & Q_c + Q_c^\top - P_c \end{bmatrix} \succ 0, \qquad (9)$$

where $\widetilde{F}_c := W_c F_c$, $\widetilde{G}_c := W_c G_c$, and $W_c := \mathrm{diag}(L_{p_{C_1}}, \dots, L_{p_{C_{n_C}}}, I, W_s)$, then the closed-loop system is δISS by setting $\begin{bmatrix} V_{\mathrm{out}_{1,c}} & V_{\mathrm{out}_{2,c}} \end{bmatrix} = L_c Q_c^{-1} E^{-1}$. $\qquad\square$

Finally, under the convergence of O, the δISS of the closed-loop system and of $\mathcal{H}$, the perfect tracking of constant signals is guaranteed by [13, Theorem 10.2.1].

3.3 Performance Optimization

The VRFT approach can be adopted to confer dynamic performances in a computationally lightweight way, provided that a dataset $u_s(k)$, $y_s(k)$, for $k = 0, \dots, N$, is available. The design problem is splitted in (i) the design of C and (ii) the design of $\mathcal{H}$, where $\mathcal{H}$ is useful to further enhance the dynamic performances. We focus on the former case and the reader is referred to [13, Sect. 10.2.4] for the latter case.

The VRFT problem consists in the minimization of

$$J_{VR}(V_{\mathrm{out}_{1,c}}, V_{\mathrm{out}_{2,c}}) = \frac{1}{N - K_0} \sum_{k=K_0}^{N-1} \left\| u_s(k) - \begin{bmatrix} V_{\mathrm{out}_{1,c}} & V_{\mathrm{out}_{2,c}} \end{bmatrix} \begin{bmatrix} \tilde{x}_c(k) \\ \tilde{u}_c(k) \end{bmatrix} \right\|^2, \qquad (10)$$

where K_0 is considered to discard the initial transient, whereas $\tilde{x}_c(k)$ and $\tilde{u}_c(k)$ depend on the *virtual reference* $\tilde{r}(k) = \mathcal{M}^{-1} y_s(k)$, as discussed in [13, Sect. 10.2.4]. Note however that we cannot, at the same time, directly minimize J_{VR} and enforce the stability constraint (9) since the free variables in inequality (9) are Q_c, L_c, and P_c. To find a unifying LMI problem, we need to define $\mathbf{U} := \begin{bmatrix} u_s(K_0) \dots u_s(N-1) \end{bmatrix}^\top$,

$$\mathbf{X} := \begin{bmatrix} \tilde{x}_c(K_0)^\top & \tilde{u}_c(K_0)^\top \\ \vdots & \vdots \\ \tilde{x}_c(N-1)^\top & \tilde{u}_c(N-1)^\top \end{bmatrix}, \quad \widetilde{\mathbf{X}} := E^\top (\mathbf{X}^\top \mathbf{X})^{-1} \mathbf{X}^\top,$$

where E is defined in Eq. (8) and the following assumption holds.

Assumption 2 Matrix $\mathbf{X}^\top \mathbf{X}$ is invertible.

The following result is proved in [13, Proposition 10.2.1].

Proposition 1 *Let Assumption 2 hold. The following optimization problem*

$$\underset{L_c \in \mathbb{R}^{m_s \times \tilde{n}}, \, \Phi_c \in \mathbb{S}^{m_s}}{\text{minimize}} \quad \mathrm{tr}(\Phi_c) \qquad (11a)$$

$$\text{subject to} \quad \begin{bmatrix} \Phi_c - \mathbf{U}^\top \widetilde{\mathbf{X}}^\top Q_c \widetilde{\mathbf{X}} \mathbf{U} + L_c \widetilde{\mathbf{X}} \mathbf{U} + \mathbf{U}^\top \widetilde{\mathbf{X}}^\top L_c^\top & L_c \\ L_c^\top & Q_c \end{bmatrix} \succcurlyeq 0 \qquad (11b)$$

is equivalent to minimize the cost function (10) if $\left[V_{\text{out}_{1,c}} \; V_{\text{out}_{2,c}} \right] = L_c Q_c^{-1} E^{-1}$ *holds and, for any scalar* $\gamma_c > 0$,

$$Q_c = \gamma_c E^{-1} \mathbf{X}^\top \mathbf{X} E^{-\top}. \tag{12}$$

Note that the constraint (12) can be possibly relaxed by defining Q_c as an optimization variable (cf. [13, Eq. (10.15)]). Also, the result in Lemma 1 can now be combined with the one in Proposition 1, leading to a unifying LMI optimization problem where both closed-loop stability and performances are taken into account (cf. [13, Sect. 10.2.5]). Finally, in [13, Sect. 10.3], the design procedure is tested on a realistic case study, i.e., a simulated model of the *pH* neutralization process.

4 Related Topics and Extensions

Below some topics and extensions related to the proposed framework are briefly outlined. Further details can be found in [13].

i. *Linear systems*—An analogous framework has been proposed to control linear systems affected by measurement noise in case of (i) polytopic uncertainty [18, 19], (ii) input saturations [20], and (iii) ellipsoidal uncertainty [21].

ii. *Robust closed-loop δISS guarantees for echo state-based control systems*—An extension of the methodology to robust design has been carried out in case of systems based on ESNs and affected by noise [13, Chap. 11].

iii. *Regional stability conditions for RNNs*—Alternative less conservative LMI conditions for regional stability have been proposed for RNNs and also applied to control design in [13, Chaps. 8 and 12].

5 Conclusions

In this Brief a novel framework for the application of VRFT to control systems described by RNNs, while providing closed-loop δISS guarantees, has been summarized. The proposed methodology, based on a novel condition for δISS of a class of RNNs, involves only computationally lightweight unifying LMI optimization problems. Possible extensions have also been outlined.

Acknowledgements The author acknowledges Prof. Marcello Farina and Dr. Alessio La Bella for their research supervision and guidance.

Competing Interests The author has no conflicts of interest to declare that are relevant to the content of this chapter.

References

1. Campi, M.C., Savaresi, S.M.: Direct nonlinear control design: the virtual reference feedback tuning (VRFT) approach. IEEE Trans. Autom. Control **51**(1), 14–27 (2006)
2. Campi, M.C., Lecchini, A., Savaresi, S.M.: Virtual Reference Feedback Tuning (VRFT): a new direct approach to the design of feedback controllers. In: 39th IEEE Conference on Decision and Control (CDC), vol. 1, pp. 623–629 (2000)
3. Sala, A., Esparza, A.: Extensions to "virtual reference feedback tuning: a direct method for the design of feedback controllers". Automatica **41**(8), 1473–1476 (2005)
4. Terzi, E., Bonetti, T., Saccani, D., Farina, M., Fagiano, L., Scattolini, R.: Learning-based predictive control of the cooling system of a large business centre. Control Eng. Pract. **97**, 104348 (2020)
5. D'Amico, W., Farina, M., Panzani, G.: Recurrent neural network controllers learned using virtual reference feedback tuning with application to an electronic throttle body. In: European Control Conference (ECC), pp. 2137–2142 (2022)
6. Angeli, D.: A Lyapunov approach to incremental stability properties. IEEE Trans. Autom. Control **47**(3), 410–421 (2002)
7. Bonassi, F., Farina, M., Scattolini, R.: Stability of discrete-time feed-forward neural networks in NARX configuration. IFAC-PapersOnLine **54**(7), 547–552 (2021)
8. Bonassi, F., Farina, M., Scattolini, R.: On the stability properties of gated recurrent units neural networks. Syst. Control Lett. **157**, 105049 (2021)
9. Bugliari Armenio, L., Terzi, E., Farina, M., Scattolini, R.: Model predictive control design for dynamical systems learned by echo state networks. IEEE Control Syst. Lett. **3**(4), 1044–1049 (2019)
10. Terzi, E., Bonassi, F., Farina, M., Scattolini, R.: Learning model predictive control with long short-term memory networks. Int. J. Robust Nonlinear Control **31**(18), 8877–8896 (2021)
11. de Souza, C., Tarbouriech, S., Girard, A.: Event-triggered neural network control for LTI systems. IEEE Control Syst. Lett. **7**, 1381–1386 (2023)
12. de Souza, C., Girard, A., Tarbouriech, S.: Event-triggered neural network control using quadratic constraints for perturbed systems. Automatica **157**, 111237 (2023)
13. D'Amico, W.: Data-based control design for linear and recurrent neural network models with stability guarantees. Ph.D. Thesis, Politecnico di Milano, Supervisor: M. Farina (2024)
14. D'Amico, W., La Bella, A., Farina, M.: An incremental input-to-state stability condition for a class of recurrent neural networks. IEEE Trans. Autom. Control **69**(4), 2221–2236 (2024)
15. Jaeger, H.: The "echo state" approach to analysing and training recurrent neural networks-with an erratum note. German National Research Center for Information Technology GMD Technical Report **148**(34), 13 (2001)
16. Sontag, E.D.: Neural nets as systems models and controllers. In: Proceedings of the Seventh Yale Workshop on Adaptive and Learning Systems, vol. 73 (1992)
17. D'Amico, W., La Bella, A., Dercole, F., Farina, M.: Data-based control design for nonlinear systems with recurrent neural network-based controllers. IFAC-PapersOnLine **56**(2), 6235–6240 (2023)
18. D'Amico, W., Farina, M.: Data-based control design for linear discrete-time systems with robust stability guarantees. In: 61st IEEE Conference on Decision and Control (CDC), pp. 1429–1434 (2022)
19. D'Amico, W., Farina, M.: Virtual reference feedback tuning for linear discrete-time systems with robust stability guarantees based on set membership. Automatica **157**, 111228 (2023)
20. D'Amico, W., Zanini, S., La Bella, A., Farina, M.: LMI-based control design with robust local stability guarantees for linear discrete-time systems with input saturations. In: European Control Conference (ECC), pp. 3490–3495 (2024)
21. D'Amico, W., Bisoffi, A., Farina, M.: Data-based control design for output-error linear discrete-time systems with probabilistic stability guarantees. IEEE Control Syst. Lett. **7**, 2035–2040 (2023)

In Silico Modelling, Analysis, and Control of Complex Diseases: Addressing Clinical Questions, Personalized Treatments, and Healthcare Management

Matteo Italia⊙ **and Fabio Dercole**⊙

Abstract Human diseases are complex and dynamic. Understanding and control-ling diseases require interdisciplinary approaches, aided by advances in digital tech-nology, data analysis, and computational power. Specifically, in his Ph.D. Thesis, Matteo Italia has developed in silico models to study cancers, Restless Legs Syn-drome (RLS), and Covid-19. The goals are to answer clinical questions, optimize treatments, and manage healthcare. For cancers, the developed models suggest that dynamic and personalized protocols can overcome drug resistance more effectively than static protocols. For neuroblastoma, the *MYCN* gene's role in treatment out-comes is explored. For melanoma, promising drug combinations are identified to overcome vemurafenib resistance. In RLS, the first mathematical model supports the hypothesis that a single neuronal generator triggers periodic leg movements, aiding disease understanding. For Covid-19, a new compartment model, including vaccina-tion policies and protection waning, emphasizes the importance of global equitable vaccine access to mitigate the pandemic. Overall, this ensemble of works highlights the importance of a systematic computational methodology in healthcare, a sort of engineered *modus operandi* that combines data analysis, systems and control, math-ematics, optimization, simulations, and coding, among others.

1 Introduction

Human diseases are often associated with complex and dynamical characteris-tics. Complexity is caused by a combination of multiple genetic, metabolic, envi-ronmental, lifestyle, and still many unknown or uncertain intertwined factors.

M. Italia (✉)
Mathematical Oncology Laboratory, Universidad de Castilla-La Mancha, Ciudad Real, Spain
e-mail: matteo.italia@uclm.es

F. Dercole
Department of Electronic, Information, and Bioengineering, Politecnico di Milano, Milano, Italy
e-mail: fabio.dercole@polimi.it

© The Author(s) 2025
S. Garatti (ed.), *Special Topics in Information Technology*,
PoliMI SpringerBriefs, https://doi.org/10.1007/978-3-031-80268-3_8

Dynamism refers to striking changes in the dynamics of some bodily functions. Most diseases show the typical structural traits of complex systems, such as positive and negative feedbacks in cellular communication, high degrees of inter- and intra-cellular connections, and interactions across different spatio-temporal scales, from molecules and fast metabolic dynamics to the entire body and slower evolutionary responses. Additionally, the related dynamical processes show the typical phenomena of complex nonlinear dynamics, such as self-sustained metabolic and neural oscillations, emergent behaviours in cell populations, such as neuronal synchronization and cancer evolutionary adaptation, wave propagation of infectious contacts, and bifurcations, for example, the transition from a constant to a dynamic optimal treatment. Complexity and dynamism are thus intertwined aspects of diseases.

Understanding and controlling the evolution of a complex disease and how to optimally treat each patient are crucial challenges that benefit the contributions of engineering, mathematics, and systems and control theory in particular. There are many open questions, especially in the applications of personalized medicine. The improvement in technology, biological data collection, effective data analysis methods, and the available computational power facilitate this research mission.

This article briefly summarizes Matteo Italia's Ph.D. Thesis in Information Technology—Systems and Control at Politecnico di Milano, Milan, Italy, supervised by Professor Fabio Dercole. This work fits into a multidisciplinary context, known as in silico medicine. The general aim is to provide modelling-informed support to healthcare. To this endeavour, Italia develops and analyzes mathematical models designed to investigate the disease under study. The models, once calibrated and validated on available data, are used to address clinical questions, personalize treatments, and optimize healthcare management. This is achieved by applying an engineered *modus operandi* that includes key steps: defining the research problem, data pre-processing and analysis, model development, calibration, validation, experiment integration, simulations, result interpretation, and analysis. This approach combines various scientific disciplines, including engineering, systems and control, mathematics, and coding. Genetic, neurological, and epidemiological diseases, specifically, cancers showing fast pharmacoresistant response, the Restless Legs Syndrome (RLS), and the Covid-19 pandemic, are studied and briefly presented as follow.

The impact of improving cancer treatments is just immense. We aim to optimally overcome drug resistance investigating cell-based (CB) and population-based (PB) evolutionary models of cancer growth under chemotherapy. Section 2 faces optimal drug scheduling in a CB theoretical framework for the first time. Section 3 presents the first PB model describing neuroblastoma optimal treatments under cyclophosphamide and vincristine. Their principal result is that it is possible to steer the evolution of cancer drug resistance, transforming it into a weakness to be exploited by optimal personalized treatments. The complex relation between the *MYCN* gene and treatment outcomes defines the *MYCN* enigma in neuroblastoma. Using a gene regulatory network, Sect. 4 discusses the impact of *MYCN* regulation on apoptosis through the *ARF-MDM2-p53* signalling pathway. Finally, Sect. 5 presents how overcoming vemurafenib resistance in melanoma by adopting the SynGeNet method.

Section 6 investigates if a single neuronal generator may trigger periodic leg movements (PLMs), a key disorder of RLS-affected subjects, rather than multiple asynchronous generators (the single-generator hypothesis). Designing and calibrating the first in silico model simulating PLMs, we support the single-generator hypothesis.

Finally, Sect. 7 studies optimal vaccination strategies against Covid-19. The main message is that stopping this pandemic requires actions to increase vaccine access, and the needed increase is less severe by adopting global equitable access.

In conclusion, Italia employs an engineered *modus operandi* to tackle complex medical problems in his Thesis. For each considered disease, new mathematical models are designed, calibrated, and validated using real data, providing valuable insights. Optimal personalized treatments are proposed for several types of cancer, a specific clinical question is addressed for Restless Legs Syndrome (RLS), and vaccination control policies are suggested to combat the Covid-19 pandemic. Section 8 resumes the key results, their impacts, and possible future developments highlighting the clinical point of view.

2 Optimal Treatments in a Cell-Based (CB) Cancer Model Suggests Switching Among Two Alternative Drugs

Many aggressive cancers remain incurable due to their rapid development of drug resistance. We develop a CB model to describe cancer growth and evolution, assuming genetic resistance to two specific chemotherapy drugs, to investigate optimal treatments. This study extends the population-based (PB) model published by Orlando and colleagues [1], which describes the evolution of a homogeneous population of cancer cells according to a fitness landscape. The landscape delineates three distinct types of trade-offs: cells are either more, less, or equally effective when developing resistance to both drugs versus specializing in resistance to a single drug.

The CB framework allows us to account for genetic heterogeneity among cells (modeling drug resistance with two continuous phenotypes), cancer cell spatial competition (affecting cancer duplication), drug diffusion dynamics, and realistic treatment protocols (with two control variables governing drug administrations).

By calibrating our model based on Orlando et al.'s assumptions, we demonstrate that protocols alternating the two drugs minimize cancer size either by the end of the treatment or at intermediate control stages. These findings significantly contrast with those obtained using the PB model, underscoring the crucial role of spatial and genetic heterogeneities in oncology. Our study represents the first effort to identify optimal treatments within a CB framework, marking a significant step toward clinical applications.

A preliminary version of this work was presented at the ECC22 conference [2], and the complete version has been published in [3].

3   Optimal Personalized Treatments in a Population-Based Model of Neuroblastoma Under Chemotherapy

Neuroblastoma is one of the most common tumors in children, with over half of high-risk patients not surviving despite multi-modal therapy. A significant challenge is the one-size-fits-all approach of induction chemotherapy (rapid COJEC), which uses fixed doses in eight two-week cycles, resulting in widely varying outcomes likely due to differences in drug resistance heterogeneity [4].

We design and calibrate a population-based (PB) model using public data to describe the evolution of neuroblastoma under treatment with vincristine and cyclophosphamide, two key components of rapid COJEC, by incorporating their pharmacokinetics. We assume that cancer cells undergo three processes: growth, mutation, and drug-induced death. The rates of these processes depend on genotype, phenotype, and environment. Cancer cells can have 3 levels of genetic resistance to each drug: none, mild, and strong, resulting in 9 different clones (genotype) within the tumor. Additionally, cancer cells can reversibly adapt (phenotype) to prolonged drug exposure at the expense of growth. We develop an optimization algorithm (genetic algorithm for global search combined with Matlab *fmincon* function for local search) to minimize tumor size prior to surgery, based on distinct pre-treatment clonal compositions representing different virtual patients.

The optimal treatments leverage the varying cytotoxic properties of the 2 drugs and the competition among different clones within the tumor, suggesting that chemotherapy can be significantly enhanced by using personalized schedules. We propose that a more comprehensive multi-modal therapy approach can be improved by integrating targeted therapies specifically directed against mutations and oncogenic pathways enriched and activated by the chemotherapeutic agents. This would require the development of a decision support system informed by emerging medical technologies, such as multi-region sequencing and liquid biopsies, to estimate patient-specific tumor heterogeneity. These results, published in [5], could serve as the groundwork for establishing such a decision support system.

4   Addressing the *MYCN* Enigma in Neuroblastoma

This work investigates the role of the *MYCN* oncogene in determining treatment outcomes for neuroblastoma. *MYCN* amplification and overexpression are critical biomarkers associated with poor prognosis. Most patients without *MYCN* amplification respond well to therapy, but some still experience unfavorable outcomes. In these cases, tumors show low *MYCN* mRNA levels but high *MYCN* protein levels, indicating protein stabilization [6]. This complex relationship between *MYCN* and treatment outcomes defines the *MYCN* enigma.

We use an in silico approach to explore the impact of *MYCN* regulation on apoptosis through the *ARF-MDM2-p53* signaling pathway. By employing a nonlinear system of ODEs based on documented gene interactions and classical gene expression modeling [7], we conduct experiments to understand the gene regulatory network dynamics. Specifically, we examine how *MYCN* affects the expression of *p53*, a tumor suppressor gene that activates apoptosis pathways. The relative expression of *p53* and *MYCN* proteins serves as an indicator of treatment success.

Model parameters are sourced from public datasets and literature. By manipulating stressors and transcription/translation rates, we identify parameter combinations that lead to favorable or unfavorable outcomes. We analyze these combinations using the Apriori algorithm for discretely distributed parameters and visualization tools for randomly distributed parameters.

Our findings show that stressors significantly impact *p53* expression, with DNA damage stress being the most crucial. This stress promotes *p53* mRNA binding to *MDM2*, enhancing *p53* translation. The production rates of *MYCN* and *p53* are primary determinants of outcomes, with their interrelationship essential for modeling the *MYCN* enigma observed clinically.

5 Searching for Effective Drug Combinations to Overcome Vemurafenib Resistance in *BRAF*-Mutant Melanoma

Melanoma has become a significant global healthcare issue primarily due to excessive ultraviolet radiation exposure. The *BRAF V600E* mutation is crucial in melanoma development, driving increased proliferation and survival. Vemurafenib, a *BRAF* inhibitor, has shown promise as a first-line treatment; however, its effectiveness diminishes over time due to the development of resistance, and no effective second-line therapies are currently available.

This study explores effective drug combinations to address vemurafenib-resistant melanoma (VRM), aiming for synergistic effects. We use the SynGeNet (Synergy from Gene expression and Network mining) methods [8] to predict effective drug combinations for VRM. SynGeNet constructs a protein-protein interaction network using deferentially expressed genes in sensitive and VRM (RNA sequencing) and the public BioGRID database, applying the belief propagation algorithm. We evaluate drug combinations based on their ability of inducing gene expression profiles exhibiting anti-correlation to the subnetwork (LINCS L1000 database) and the centrality of drug targets (DrugBank and STITCH databases) within the network. The most promising combination identified is sorafenib and pioglitazone.

In vitro tests, including viability assessments and synergy analysis, guide the determination of optimal drug concentrations. These tests confirm the efficacy of the combination, which should be further validated before clinical applications.

6 A Model-Informed Answer to the Single-Generator Hypothesis Triggering Periodic Leg Movements (PLMs)

Here, we address a clinically-driven question about PLMs, the principal disorder of people affected by RLS. The goal is to employ numerical simulations to assess whether the complex process involving two (or more) interacting spinal or supraspinal structures responsible for generating PLMs can be effectively modeled using a single-generator approach (single generator hypothesis). Thus, we develop the first phenomenological model generating virtual or in silico PLMs.

We simplify the physiology and assumed that each leg is controlled by a single motor neuron, representative of the central nervous system (CNS) complex pathways, determining contractions of the leg muscles. The two neurons are modeled with integrate-and-fire neuron models. Each neuron receives excitatory/inhibitory inputs from the CNS physiological activity, a proportion of which equally affect both legs, while the remaining fraction is leg-specific. We calibrate this physiological activity using the control dataset (healthy subjects). For subjects showing significant PLMs, we introduce a periodic excitatory input, calibrated on RLS dataset (pathological subjects), that influence both legs representing the pathological activity.

Despite simplifying assumptions, our model-simulated data closely resemble the real polysomnographic data, displaying minimal significant differences. This provides compelling preliminary evidence supporting the single generator hypothesis for PLMs. Future model extensions will strive to develop a diagnostic and therapeutic tool that can assist healthcare specialists in making realistic forecasts and in cross-correlating and clustering data with other patient-specific information.

A preliminary and modeling oriented version of this work has been presented at the CIBB2021 conference [9], and the complete article published in [10].

7 Optimal Vaccination Strategy Suggests Enhancing Global Equity and Access to Covid-19 Vaccines

We adopt a model-informed approach to argue that Global Equitable Access (GEA) to Covid-19 vaccines is crucial to ending the pandemic. We develop a new epidemiological compartmental model, calibrated with public data (Our World in Data database), integrating the principal aspects of the Covid-19 pandemic: susceptible, exposed, presymptomatic, symptomatic and asymptomatic infected, quarantined, hospitalized, recovered, and death subjects; contact tracing; containment measures; vaccination campaign. Two interconnected macro-nodes describe the world, high-income (HI) and middle-to-low-income (MLI) countries, with the same epidemiological model but a different vaccine availability.

Our findings indicate that merely redistributing the currently available vaccines equitably is insufficient to stop the pandemic. However, a significant increase in vaccine production (access to vaccines) could end the pandemic within a year of vaccination and save millions of lives in MLI countries. We also explore the relationship between vaccine access and distribution among HI and MLI countries, showing that the required access to vaccines to stop the pandemic gets reduced the more global equity is used.

To estimate the socio-economic impacts, we compare simulations of the current scenario–where the virus remains potentially harmful and is likely to become endemic–with a hypothetical but realizable scenario, where the used vaccine rollouts are the peaks achieved during the 2021–2022 vaccination campaign. Our analysis shows that the global savings on vaccines in the selected scenario surpass the five-year profit of the rights holders under the current situation, justifying compensation mechanisms for necessary licensing agreements. The encouraging news is that the benefits of the selected scenario are still significant if implemented now.

The preliminary version of this work has been presented at the COMPENG 2022 conference [11], and the complete version published in [12].

8 Conclusion

Matteo Italia's Ph.D. Thesis focuses on in silico medicine, an interdisciplinary field gaining attention for its use of computational methods to advance medical understanding and treatment, through the lens of personalized medicine. The in silico term represents a paradigm shift in the research and applications of medicine, bringing significant transformative effects. The constructive results, achieved applying Italia's engineered *modus operandi*, are the core contributions of his Thesis. In the following, the key results, their impacts, and possible future developments are resumed, underlining both the research and clinical points of view, such as the understanding of physiological and pathological mechanisms, the evolution of a disease, treatments focusing on optimal and personalized strategies, and healthcare management.

Cancers are highly complex and adaptive because of high mutation rates. Indeed, chemotherapy usually kills cancer cells effectively initially, but then cancers achieve resistance. We investigate how to overcome drug resistance and treatment failures.

Designing, calibrating, and validating valuable predictive models, all the desired treatment strategies can be virtually implemented, tested, and evaluated before applying the optimal treatments to real patients. Specifically, in the theoretical framework presented in Sect. 2, we found that it is possible to exploit cancer drug resistance, steering cancer evolution towards evolutionary traps by exploiting evolutionary double binds. The optimal administration strategies are dynamic protocols that alternate between two alternative drugs (bang-bang control), obtaining substantial reductions in final tumor sizes. Moreover, Sect. 3 shows that when administering vincristine and cyclophosphamide to neuroblastoma patients, the optimal personalized treatments are based on general evolutionary principles that exploit the clonal compositions of the tumor. We discover that starting the treatment with only one drug and adding the other later, and using both drugs from the beginning but shortening the number of treatment cycles are optimal strategies for specific initial cancer compositions mimicking specific patients. The personalized optimal treatments highly outcompete standard one-size-fits-all protocols.

Section 4 addresses the *MYCN* enigma (the complex relationship between the *MYCN* gene and the treatment outcomes) in neuroblastoma treatments exploiting a calibrated gene regulatory network. Using the relative expression of *p53* (tumor suppressor gene) and *MYCN* (oncogene) as an indicator of treatment outcome, we found in silico outcomes in agreement with clinical outcomes. Moreover, we postulated that therapeutic strategies aimed at augmenting *p53* production, diminishing *MYCN* production, and facilitating the binding of the *p53* mRNA and the *MDM2* protein hold great promise to be combined with standard treatments for enhancing neuroblastoma treatment outcomes.

Analysing gene expressions and protein-protein interactions in Sect. 5, we in silico predict pioglitazone and sorafenib as effective treatment against melanoma resistant to vemurafenib, confirmed by preliminary in vitro tests. If the drug combination is approved in clinical trials, there will be available an effective second-line therapy, significantly improving melanoma treatment outcomes.

The models of Sects. 2 and 3 can be enhanced by incorporating factors like metabolism, microenvironment, multi-drug resistance, and clinical realism. For personalized medicine, patient-specific parameters such as tumor heterogeneity, drug absorption, and cancer cell growth can be identified, allowing to virtually determine the best personalized treatments. This requires a decision support system that includes all approved drugs, relevant mutations, oncogenic pathways, and the clonal composition of tumors determined through multi-region sequencing and liquid biopsies. This system will optimize therapy by customizing chemotherapy schedules and integrating other treatments like surgery and immunotherapy. Similarly, the gene regulatory network of Sect. 4 can be personalized to predict patient-specific outcomes and adjust medical interventions. Additionally, the methodology from Sect. 5 can serve as a foundation for personalized treatment decision support systems for other cancers at various stages. Using patient RNA sequencing data, patient-specific disease networks can predict effective personalized drugs. These predicted drugs can then be tested in vitro with patient-derived cancer cells before being administered to the patient.

Restless legs syndrome (RLS) is a common yet often overlooked condition marked by periodic leg movements (PLMs) during sleep, leading to disrupted sleep. Section 6 investigates the hypothesis that a single neuronal generator causes PLMs. The developed model, calibrated with clinical data, supports the hypothesis that a single generator could trigger PLMs. This understanding is vital for treatment decisions. The model provides a foundation for future research, potentially reducing costs, avoiding animal testing, simulating drug effects, and enhancing knowledge of RLS and PLMs by introducing new indicators and considering patient metadata.

Covid-19 has profoundly impacted the world, with vaccine access disparities between high-income (HI) and middle-to-low-income (MLI) countries hindering pandemic control. Section 7 presents a model showing that Global Equitable Access (GEA) to vaccines is crucial. The study examines vaccination strategies by dividing the world into HI and MLI countries and varying vaccine availability and distribution.

When vaccine access is limited, prioritizing HI countries helps control infections. With moderate access, focusing on MLI countries becomes optimal. When access is sufficient, GEA is the best strategy. Enhancing vaccine production and equitable distribution is essential, although socio-economic disparities make perfect equity challenging. The model shows that achieving GEA reduces the severity of required production increases to stop the pandemic. These strategies offer significant socio-economic benefits, informing current and future pandemic responses and aiding policymakers in optimizing vaccination campaigns and resource management.

Competing Interests The authors have no conflicts of interest to declare that are relevant to the content of this chapter.

References

1. Orlando, P.A., Gatenby, R.A., Brown, J.S.: Cancer treatment as a game: integrating evolutionary game theory into the optimal control of chemotherapy. Physical Biology (2012). https://doi.org/10.1088/1478-3975/9/6/065007
2. Italia, M., Dercole, F., (2022) Optimal control of two cytotoxic drug maximum tolerated dose steers and exploits cancer adaptive resistance in a cell-based framework.: European Control Conference. ECC **2022**,(2022). https://doi.org/10.23919/ECC55457.2022.9838039
3. Italia, M., Dercole, F., Lucchetti, R.: Optimal chemotherapy counteracts cancer adaptive resistance in a cell-based, spatially-extended, evolutionary model. Physical Biology (2022). https://doi.org/10.1088/1478-3975/ac509c
4. Smith, V., Foster, J.: High-Risk Neuroblastoma Treatment Review. Children (2018). https://doi.org/10.3390/children5090114
5. Italia, M., Wertheim, K.Y., Taschner-Mandl, S., Walker, D., Dercole, F.: Mathematical Model of Clonal Evolution Proposes a Personalised Multi-Modal Therapy for High-Risk Neuroblastoma. Cancers (2023). https://doi.org/10.3390/cancers15071986
6. Valentijn, L., Koster, J., Haneveld, F., Aissa, R., Sluis, P., Broekmans, M., Molenaar, J., van Nes, J., Versteeg, R.: Functional MYCN signature predicts outcome of Neuroblastoma irrespective of MYCN amplification. Proceedings of the National Academy of Sciences of the United States of America (2012). https://doi.org/10.1073/pnas.1208215109
7. Ingalls, B.: Mathematical Modeling in Systems Biology. MIT Press (2013)
8. Regan-Fendt, K.E., Xu, J., DiVincenzo, M., Duggan, M.C., Shakya, R., Na, R., Carson, W.E., III., Philip, R.O., Payne, P.R.O., Li, F.: Synergy from gene expression and network mining (SynGeNet) method predicts synergistic drug combinations for diverse melanoma genomic subtypes. Npj System Biology Application (2019). https://doi.org/10.1038/s41540-019-0085-4
9. Italia, M., Danani, A., Dercole, F., Ferri, R., Manconi, M.: The First in-silico Model of Leg Movement Activity During Sleep. Lecture Notes in Computer Science (including subseries Lecture Notes in Artificial Intelligence and Lecture Notes in Bioinformatics) (2022). https://doi.org/10.1007/978-3-031-20837-9_4
10. Italia, M., Danani, A., Dercole, F., Ferri, R., Manconi, M.: A calibrated model with a single-generator simulating polysomnographically recorded periodic leg movements. Journal of Sleep Research (2022). https://doi.org/10.1111/jsr.13567

11. Italia M, Della Rossa F, Dercole F (2022) A mathematical analysis of the socio-economic impacts of a patent waiver on COVID-19 vaccines: IEEE Workshop on Complexity in Engineering. COMPENG (2022). https://doi.org/10.1109/COMPENG50184.2022.9905430
12. Italia, M., Della Rossa, F., Dercole, F.: Model-informed health and socio-economic benefits of enhancing global equity and access to Covid-19 vaccines. Scientific Reports (2023). https://doi.org/10.1038/s41598-023-48465-y

Control of Large-Scale MLD Systems via Multi-agent Reformulation and Decentralized Optimization

Lucrezia Manieri ⓘ

Abstract Motivated by the challenges emerging in the energy sector, this brief presents a comprehensive framework for the optimal operation of large-scale Mixed Logical Dynamical (MLD) systems modeling engineering systems characterized by interleaved physical, logical, and digital components and subject to operational constraints. When the performance index is linear, the problem translates into a Mixed Integer Linear Program (MILP) that is NP-hard and becomes prohibitive as the size of the system increases. In the case of multi-agent systems that are constraint-coupled, decentralized schemes with provable feasibility guarantees are introduced to recover computational tractability by reducing the global MILP to multiple smaller ones that are iteratively solved in parallel. If the matrix modeling the MILP constraints is sparse, a method is proposed to possibly recover a hidden constraint-coupled multi-agent structure to which the decentralized resolution schemes can then be applied. Finally, multi-agent constraint-coupled MILPs with uncertain local constraints are considered and probabilistic feasibility guarantees are derived for their data-based decentralized solution. The framework is tested on an application in the energy sector concerning the provision of ancillary services to the power distribution grid.

1 Introduction

The global thrust for sustainable electrification, where fossil fuel-based technologies and processes are replaced by electrically-powered equivalents with electricity produced from renewable energy sources, is driving unprecedented changes in the energy sector, calling for the direct involvement of end-users and the aggregation of distributed energy sources to support the grid operators [1, 2]. This results in an increase in both scale and complexity of the involved engineering systems, that are evolving into large agglomerates of interconnected physical and digital components.

L. Manieri (✉)
Dipartimento di Elettronica, Informazione e Bioingegneria, Politecnico di Milano, Via Giuseppe Ponzio, 34/35, 20133 Milano, Italy
e-mail: lucrezia.manieri@polimi.it

© The Author(s) 2025
S. Garatti (ed.), *Special Topics in Information Technology*,
PoliMI SpringerBriefs, https://doi.org/10.1007/978-3-031-80268-3_9

In this brief, we focus on those engineering systems that can be modeled as Mixed Logical Dynamical (MLD) systems [3]. MLD systems are described via linear equality and inequality constraints involving both continuous and binary variables and can model a wide range of systems with interleaved physical, logical, and digital components, subject to constraints of various nature, including propositional logic constraints. When the performance index is also linear, the optimal operation of an MLD system translates into a Mixed-Integer Linear Program (MILP). As the size of the MLD system increases, the combinatorial complexity of the resulting mixed-integer optimization program grows exponentially and finding a solution becomes challenging, if not prohibitive. The presence of uncertainty affecting the system adds further complexity to the problem. In the quite common case when the uncertainty is known only from data, a solution with guaranteed generalization properties from seen to not yet seen uncertainty instances is needed.

This brief outlines the optimization framework proposed in [4] to address these complexity challenges. The framework includes a decomposition strategy to reveal a possibly hidden multi-agent structure of an MLD system, decentralized resolution schemes that apply to a multi-agent structured MLD system and recover computational tractability by decomposing the corresponding MILP into smaller problems (one per agent), and a data-based privacy-preserving approach to decentralized multi-agent constraint-coupled MILP resolution with probabilistic feasibility guarantees.

Such an optimization framework is mainly motivated by emerging challenges in the energy sector and can capture a wide range of applications such as optimal charging scheduling of plug-in electric vehicles, economic dispatch problems, and provision of ancillary services in smart energy systems. We refer the interested reader to [4–6] for a detailed description of these applications and numerical simulation examples showcasing the effectiveness of the proposed framework.

The remainder of the brief is structured as follows. Section 2 shows how the optimal operation of an MLD system can be formulated as an MILP and introduces a class of multi-agent MILPs whose structure can be leveraged to recover computational tractability. Section 3 outlines a strategy to disclose the hidden multi-agent structure (if any) of a linearly constrained optimization program via manipulation of its constraint matrix. Section 4 proposes two decentralized resolution schemes for constraint-coupled multi-agent MILPs and a further one for non-convex—but not necessarily mixed-integer linear—optimization problems that are characterized by a scalar complicating constraint. Section 5 addresses constraint-coupled MILPs in which the agents' local constraints are affected by uncertainty that is only known through a set of *private* data that cannot be shared with other agents. Probabilistic feasibility guarantees are derived for a (possibly sub-optimal) solution obtained via decentralized schemes that preserve the privacy of the local information. Section 6 discusses the application of the proposed framework to balancing services provision in smart energy systems. Finally, Sect. 7 concludes this brief.

2 Large-Scale MLD Systems with a Hidden Multi-Agent Structure

A Mixed Logical Dynamical (MLD) system is described by the following equations

$$\xi(t+1) = F_t \xi(t) + V_{u,t} u(t) + V_{\sigma,t} \sigma(t) \tag{1a}$$

$$M_{\sigma,t} \sigma(t) \le M_{u,t} u(t) + M_{\xi,t} \xi(t) + M_t \tag{1b}$$

where the state, input, and auxiliary vectors ξ, u, and σ have both continuous and discrete components. The inequality in (1b) should be interpreted component-wise, and collects all linear operational constraints and the reformulation of piece-wise affine dynamics and logical conditions by means of some suitably defined auxiliary variables entering σ. When addressing the optimal control of an MLD system over some finite time horizon $\mathcal{T} \subset \mathbb{N}$, Eq. (1) are integrated into a mathematical program where some cost function is minimized with respect to the input sequence. If the performance index is a linear function of the state, input and auxiliary variables, then the resulting optimization problem is a Mixed-Integer Linear Program (MILP)

$$\min_{x} \quad c^{\top} x \tag{2a}$$

$$\text{subject to:} \quad Ax \le b \tag{2b}$$

$$x \in \mathbb{R}^{n_c} \times \mathbb{Z}^{n_d} \tag{2c}$$

where x is the decision vector with n_c continuous and n_d discrete components that collect the input and auxiliary variables along the horizon $\mathcal{T}$. Vector c in (2a) defines the cost function, while the inequality (2b) with matrix A and vector b represents in compact form the MLD constraints with the state expressed as a function of the decision variables x and the initial state by unrolling the MLD dynamics over $\mathcal{T}$.

All values of x satisfying the constraints (2b) and (2c) are said to be *feasible* solutions of the MILP (2) and they define its mixed-integer *feasibility set* $X = \{x \in \mathbb{R}^{n_c} \times \mathbb{Z}^{n_d} : Ax \le b\}$. A feasible $x^\star \in X$ that minimizes the objective function, i.e., such that $c^{\top} x^\star \le c^{\top} x$, $\forall x \in X$, is an *optimal* solution of the MILP.

Due to the presence of discrete decision variables, MILPs have an intrinsic combinatorial complexity that grows exponentially with n_d and can make their resolution prohibitive. Depending on the problem at hand, problem-specific heuristics and relaxation schemes can be exploited to compute a feasible—although sub-optimal—solution. This is the case for constraint-coupled multi-agent MILPs of the form

$$\min_{x_1,\ldots,x_m} \quad \sum_{i=1}^{m} c_i^{\top} x_i \tag{3a}$$

$$\text{subject to:} \quad \sum_{i=1}^{m} E_i x_i \le f \tag{3b}$$

$$x_i \in X_i, \ i = 1, \ldots, m \tag{3c}$$

where x_i denotes the decision vector of agent i with $n_{c,i}$ continuous components and $n_{d,i}$ discrete ones. The local mixed-integer sets X_i, $i = 1, \ldots, m$, are defined as $X_i = \{x_i \in \mathbb{R}^{n_{c_i}} \times \mathbb{Z}^{n_{d_i}} : D_i x_i \leq d_i\}$ with matrix D_i and vector d_i of suitable dimensions. Inequality (3b) collects p coupling constraints.

Duality-based resolution methods have been proposed to decompose problems with the multi-agent structure in (3) into smaller MILPs (one per agent) and distribute the computations to recover tractability. They include the methods in [7, 8] as well as those proposed in [4] and outlined in Sect. 4. Problems in the form (3) naturally arise when addressing the optimal control of a system composed of multiple units, each one modeled as an MLD system and interacting with the other units because of some shared resources. In some cases, however, the multi-agent structure of a system may not be evident during the modeling phase. As a result, units that could be modeled as separate agents may be embedded in the same MLD system, ultimately leading to a seemingly non-separable MILP. For this reason, the optimization framework developed in this brief includes a decomposition strategy to disclose the (possibly hidden) multi-agent structure of an MILP.

3 Decomposition Strategy

The constraint-coupled multi-agent MILP in (3) can be recast as a generic MILP in the form of (2) by collecting all local decision variables and cost coefficients in the vectors $x^\top = [x_1^\top \cdots x_m^\top]$ and $c^\top = [c_1^\top \cdots c_m^\top]$ and by expressing all constraints in the compact form (2b) with

$$
A = \begin{bmatrix} D_1 & 0 & \cdots & 0 \\ 0 & D_2 & \cdots & 0 \\ \vdots & \vdots & \ddots & \vdots \\ 0 & 0 & \cdots & D_m \\ E_1 & E_2 & \cdots & E_m \end{bmatrix} \qquad b = \begin{bmatrix} d_1 \\ d_2 \\ \vdots \\ d_m \\ f \end{bmatrix}, \tag{4}
$$

where each block on the main diagonal of A defines the agents' local variables and constraints, while the horizontal border at the bottom represents the coupling constraints. Such a structure is called *singly-bordered block-diagonal*. When the optimization program also comprises variables that are not associated with a specific agent and are involved both in most of the agents' local constraints and coupling constraints, the constraint matrix A assumes a *doubly-bordered block-diagonal* structure characterized by an additional vertical border.

If an MILP (2) has a sparse constraint matrix A, one can search for a permutation of its rows and columns that brings it into a singly- or doubly-bordered block-diagonal form, thus recovering an equivalent multi-agent formulation. In [9], a strategy is proposed to bring the (sparse) constraint matrix A of an MILP into a singly-bordered block-diagonal form, while minimizing the number of coupling constraints and evenly distributing the discrete variables among the agents, thus balancing the complexity of their local MILPs. Differently from [9], in [4] we propose a strategy to seek both a singly-bordered and a doubly-bordered permuted version of matrix A. The problem is recast as a graph-partitioning problem on a suitably weighted *column-vertex* graph representation of A with each node coinciding with a decision variable—and, thus, a column of the matrix—and with edges connecting variables that are coupled by one or more constraints. The strength of the connection between two variables is coded by the weight associated with the edge connecting them, which is suitably chosen so as to reflect the sparsity structure of the matrix. If the (rescaled) weights are interpreted as transition probabilities of a Markov chain evolving on the graph, the set of nodes can be partitioned based on the similarity of the evolution of the probability distributions obtained by initializing the Markov chain at each node. The obtained partition can be then interpreted as a permutation of the rows and columns of the matrix, to identify the blocks on the diagonal and the borders. The best singly- and doubly-bordered reformulations are selected based on a user-defined criterion accounting for specific requirements on the permuted matrix (e.g., minimum size of the borders, maximum number of agents in the reformulation or a combination of the two criteria).

Numerical simulations show that the proposed decomposition strategy is more effective in disclosing singly-bordered structures than the one in [9] and is also able to disclose doubly-bordered reformulations, provided that columns in the (hidden) border are sufficiently dense. Further effort is needed to address the decomposition of constraint matrices that do not admit a block-angular form but can still be reduced to it by neglecting some constraints between decision variables of different agents.

4   Resolution Schemes for Constraint-Coupled Multi-agent MILPs

In this section, we address the resolution of large-scale MILPs with the constraint-coupled multi-agent structure in (3), possibly recovered via the decomposition strategy in Sect. 3. The presence of the coupling constraints hampers the decomposition of the problem into separate lower-dimensional MILPs, one per agent, and makes the resolution challenging. To recover computational tractability, the coupling constraints can be lifted to the cost function and penalized through a vector λ of non-negative weights called *Lagrange multipliers*. The minimum of the resulting *Lagrangian relaxation* provides a lower bound on the optimal cost of the original problem (see e.g. [10, Sect. 5.1.3]). The best lower bound is obtained by solving the *dual problem*

$$\max_{\lambda \geq 0} \min_{\{x_i \in X_i\}_{i=1}^m} \sum_{i=1}^m c_i^\top x_i + \lambda^\top \left(\sum_{i=1}^m E_i x_i - f \right) = \max_{\lambda \geq 0} -\lambda^\top f + \sum_{i=1}^m \min_{x_i \in X_i} (c_i^\top + \lambda^\top E_i) x_i , \quad (5)$$

where the inner minimization can be decoupled into m smaller MILPs of the form

$$x_i(\lambda) \in \arg\min_{x_i \in X_i}(c_i^\top + \lambda^\top E_i)x_i. \tag{6}$$

In addition, since the dual function $\varphi(\lambda) = -\lambda^\top f + \min_{x_i \in X_i}(c_i^\top + \lambda^\top E_i)x_i$ is concave in λ, problem (4) is convex despite the original MILP (3) being non-convex (see e.g. [10, Sect. 5.1.2]), and an optimal $\lambda^\star$ can be efficiently computed via a decentralized implementation of the well-established sub-gradient algorithm. Unfortunately, the recovered primal solution $x(\lambda^\star) = [x_1(\lambda^\star)^\top \cdots x_m(\lambda^\star)^\top]^\top$, with $x_i(\lambda^\star)$ given in (6) with $\lambda = \lambda^\star$, $i = 1, \ldots, m$, may not satisfy the dualized coupling constraints since they are not directly enforced in (6). To overcome such an issue, recent works [7, 8] propose a strategy that guarantees feasibility of the primal solution with respect to the coupling constraint by introducing a fictitious tightening, as explained next.

Denote with ρ a tightening vector such that $\rho \geq 0$, and replace the coupling constraint (3b) with its tightened counterpart $\sum_{i=1}^m E_i x_i \leq f - \rho$. In addition, let $\lambda_\rho^\star$ be the optimal solution of the dual of the tightened version of (3) for a generic ρ and denote with $x(\lambda_\rho^\star)$ the primal solution composed by the sub-vectors $x_i(\lambda_\rho^\star)$ defined in (6) with $\lambda = \lambda_\rho^\star$. If ρ is chosen so as to exceed the constraint violation of the recovered primal solution, i.e., $\rho \geq \sum_{i=1}^m E_i x_i(\lambda_\rho^\star) - f$, then, $x(\lambda_\rho^\star)$ is feasible for the original coupling constraint. Such a condition, however, does not provide an explicit expression for the tightening as it entails solving an optimization problem parametrized in ρ.

In [7], a worst-case $\tilde{\rho}$ is computed based on all admissible solutions of (3). The obtained solution $x(\lambda_{\tilde{\rho}}^\star)$ is granted to satisfy the coupling constraints but typically has a poor performance since its cost degradation grows with the infinity norm of the tightening $|\tilde{\rho}|_\infty$. For this reason, in [8], conservativeness is reduced by selecting the tightening in an adaptive fashion, ensuring the aforementioned condition on ρ is eventually met for a value of ρ that is non-worse than $\tilde{\rho}$ and that, thus, yields a solution attaining a smaller cost. The update is integrated into a decentralized scheme in which the agents repeatedly compute candidate primal solutions and share their impact on the coupling constraints with a central unit, which, in turn, updates the tightening vector ρ at each iteration based on the explored primal solutions.

In [4], we introduce two novel decentralized schemes that further reduce conservativeness by changing the update strategy of the tightening ρ. The first one, originally presented in [5], progressively increases the tightening based on the violation of the coupling constraint attained by the tentative primal solution $x(\lambda_\rho^\star)$ computed throughout the iterations. The second, introduced in [11], allows the tightening to both increase and decrease, depending on the mismatch between $x(\lambda_\rho^\star)$ and the solution of a relaxed version of the original problem associated with the current tightening. We refer the reader to [4, 5, 11] for a detailed description of the algorithms and the demonstration of their superiority compared with state-of-the-art resolution schemes [7, 8].

For the first algorithm, we prove convergence to a feasible primal solution in a finite number of iterations and derive a bound on the expected performance degradation, [4, Sect. 4.5.1]. Characterizing the properties of the second algorithm is more challenging due to the non-monotonic nature of the tightening update. Convergence in a finite number of iterations to a feasible primal solution is proven only for the case of a single coupling constraint by exploiting the fact that the procedure boils down to the well-established sub-gradient algorithm, [4, Sect. 4.5.2].

We then focus on non-convex programs characterized by one *complicating constraint*, i.e., a constraint such that its penalization in the cost would make the problem easier to solve. This includes but is not limited to multi-agent MILPs with a scalar coupling constraint, which hampers the decomposition of the problem and makes its resolution challenging. In [4], we propose a resolution strategy that resorts to duality theory by lifting the complicating constraint to the cost function and using a single Lagrange multiplier. Starting from the observation that solving the dual problem amounts to finding the zeros of the sub-differential of the dual function, the algorithm leverages the monotonicity of the sub-differential to solve the dual problem via bisection while generating a sequence of feasible primal solutions with non-degrading performance, in terms of cost. The resulting Dual Bisection (DualBi) algorithm is shown to provide a feasible primal solution in finite time, with a performance that progressively improves throughout iterations. When it is applied to multi-agent MILPs with a single coupling constraint, DualBi admits a decentralized implementation where agents update a portion of the candidate primal solution at each iteration, whereas a central unit performs the bisection.

Numerical simulations on economic dispatch problems with quadratic cost function and disjunctive constraints show the efficacy of the algorithm, which systematically computes feasible primal solutions with a close-to-zero optimality gap [6].

5 Dealing with Uncertainty

Finally, we consider multi-agent constraint-coupled MILPs where the agents' local constraints are affected by uncertainty, which is known to the agents through *private data* that cannot be shared with others. A data-based optimization problem where constraints are imposed by each agent on its own dataset is introduced, and the generalization properties of its solution are studied.

The scenario approach [12] provides an a-priori bound on the number of data needed to guarantee that a data-based optimal solution satisfies the constraints with a certain (high) probability. However, it requires the problem to be convex. As for the MILP case, a-priori bounds exist, but they either hold only for optimal solutions of the data-based optimization program [13] or for a feasible solution computed by the agents while sharing their datasets [14].

By exploiting tools from statistical learning theory, we derive probabilistic feasibility guarantees for a feasible solution of the data-based optimization problem. The obtained bound on the number of data is shown to depend on the complexity of the uncertain constraint set. Differently from [13], the provided guarantees also hold for sub-optimal solutions obtained via the decentralized schemes presented in

Sect. 4 that preserve the privacy of the local information of the agents. We refer the
reader to [15] for more details.

6 Application to Smart Energy Systems

The proposed framework can be applied to the problem of balancing services provision in smart energy systems described in [1]. We consider a pool of m prosumers
coordinated by an Energy Service Provider (ESP) so as to provide a certain overall
power deviation from a pre-agreed baseline upon request by the Transmission System Operator (TSO). The ESP must coordinate all prosumers so as to timely meet the
request within a certain tolerance, while minimizing operating costs. The pool can
be modeled as a multi-agent MLD system and the coordination problem formulated
as an MILP in the form of (2). As shown in [4], due to some modeling choices prescribed by the MLD formalism, the resulting constraint matrix A is sparse, yet it does
not exhibit a singly-bordered block-diagonal structure. The decomposition strategy
outlined in Sect. 3 proves effective in disclosing the hidden constraint-coupled multi-
agent structure of the problem, with a number of blocks that exceeds the number m of
prosumers. The resulting (structured) MILP is efficiently solved by the decentralized
schemes in Sect. 4 yielding close-to-optimal solutions even for instances with more
than $m = 200$ prosumers, as shown in Fig. 1 taken from [4]. In contrast, state-of-the-
art competitors [7, 8] prove overly conservative for this specific problem and cannot
provide a feasible solution even for $m = 50$.

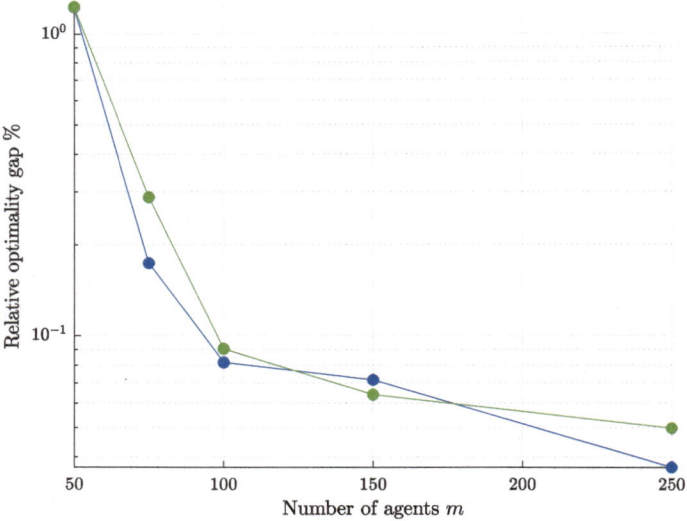

Fig. 1 Relative optimality gap (%) obtained by the decentralized algorithms with constraint tightening in Sect. 4 as a function of the size m of the pool

7   Conclusions

In this brief, we propose a computational framework to address the optimal operation of large-scale Mixed Logical Dynamical (MLD) systems with a control cost that is linear, thus resulting in a Mixed Integer Linear Program (MILP). The framework comprises decentralized resolution schemes for MILPs with a multi-agent constraint-coupled structure, including a duality-based bisection scheme that applies to more general non-convex optimization problems but with only one complicating constraint; a decomposition strategy to possibly reduce to such a structure a large size MILP; and a probabilistic certificate of feasibility for the data-based decentralized solution to a multi-agent constraint-coupled MILP when local constraints are affected by uncertainty. Further work is needed to extend the duality-based bisection scheme to the case of multiple constraints and improve the bounds for the probabilistic certificates.

References

1. La Bella, A., Falsone, A., Ioli, D., Prandini, M., Scattolini, R.: A mixed-integer distributed approach to prosumers aggregation for providing balancing services. Int. J. Electr. Power Energy Syst. **133**, 107228 (2021)
2. Askeland, M., Backe, S., Bjarghov, S., Korpås, M.: Helping end-users help each other: coordinating development and operation of distributed resources through local power markets and grid tariffs. Energy Econ. **94**, 105065 (2021)
3. Bemporad, A., Morari, M.: Control of systems integrating logic, dynamics, and constraints. Automatica **35**(3), 407–427 (1999)
4. Manieri, L.: Control of large-scale MLD systems via multi-agent reformulation and decentralized optimization. Ph.D. dissertation, Politecnico di Milano, March 2024
5. Manieri, L., Falsone, A., Prandini, M.: Handling complexity in large-scale cyber-physical systems through distributed computation. In: Computation-Aware Algorithmic Design for Cyber-Physical Systems, pp. 55–78. Springer (2023)
6. Manieri, L., Falsone, A., Prandini, M.: A dual bisection approach to economic dispatch of generators with prohibited operating zones. IEEE Control Syst. Lett. 1–1 (2024)
7. Vujanic, R., Esfahani, P.M., Goulart, P.J., Mariéthoz, S., Morari, M.: A decomposition method for large scale MILPs, with performance guarantees and a power system application. Automatica **67**, 144–156 (2016)
8. Falsone, A., Margellos, K., Prandini, M.: A decentralized approach to multi-agent MILPs: finite-time feasibility and performance guarantees. Automatica **103**, 141–150 (2019)
9. Manieri, L., Falsone, A., Prandini, M.: Hyper-graph partitioning for a multi-agent reformulation of large-scale MILPs. IEEE Control Syst. Lett. **6**, 1346–1351 (2021)
10. Boyd, S.P., Vandenberghe, L.: Convex Optimization. Cambridge University Press (2004)
11. Manieri, L., Falsone, A., Prandini, M.: A novel decentralized approach to large-scale multi-agent MILPs. IFAC-PapersOnLine **56**(2), 5919–5924 (2023)
12. Campi, M., Garatti, S., Prandini, M.: The scenario approach for system and control design. Elsevier, Ann. Rev. Control **33**, 149–157 (2009)

13. Esfahani, P.M., Sutter, T., Lygeros, J.: Performance bounds for the scenario approach and an extension to a class of non-convex programs. IEEE Trans. Automatic Control **60**(1), 46–58 (2014)
14. Falsone, A., Margellos, K., Prandini, M., Garatti, S.: A scenario-based approach to multi-agent optimization with distributed information. IFAC-PapersOnLine **53**(2), 20–25 (2020)
15. Manieri, L., Falsone, A., Prandini, M.: Probabilistic feasibility in data-driven multi-agent non-convex optimization. Ann. Rev. Control **56**, 100925 (2023)

Telecommunications

Forensic Analysis of Satellite Imagery: Challenges and Solutions

Edoardo Daniele Cannas

Abstract Satellite images are now a widespread asset that is easily obtainable on the Web. Many portals offer these images for free, and their role in sensitive applications, such as natural disaster response, intelligence, and military, is becoming paramount. For these reasons, satellite images can be a target for malicious manipulations. Multimedia Forensics (MMF) is the discipline concerned with assessing the authenticity of multimedia data. Satellite images pose new challenges to MMF due to (i) being an inherently multimodal data asset, with some of its modalities, like Synthetic Aperture Radar (SAR) signals, which the community has never investigated; (ii) having a complex processing pipeline where forensic traces are challenging to model. In this chapter, we tackle the forensic analysis of satellite images, namely panchromatic and SAR imagery, and propose using Convolutional Neural Networks (CNNs) to extract forensic information. In particular, we will consider the problems of source attribution and image splicing localization. In both situations, CNNs prove effective and more performant than techniques developed for standard digital pictures, substantiating the need for forensic tools tailored to remote sensing data.

1 Introduction

Internet-based communications have expanded to unprecedented dimensions in recent years, and nowadays, the Web offers modalities of data beyond text, images, and videos. For example, satellite imagery, freely accessible through portals like the Copernicus Open Access Hub [10] and the U.S. Geological Survey Landsat [15], has become increasingly widespread. Such images have incredible value for various applications, ranging from Earth surface monitoring [12] to intelligence applications, e.g., investigating the ongoing fighting in war zones [9].

However, this accessibility allows malicious actors to manipulate data, and unusual data modalities, like satellite images, are also possible targets. Indeed, public

E. D. Cannas (✉)
Politecnico di Milano, Dipartimento di Elettronica, Informazione, e Bioingegneria (DEIB), Via Ponzio 34/5, 20133 Milano, Italy
e-mail: edoardodaniele.cannas@polimi.it

S. Garatti (ed.), *Special Topics in Information Technology*,
PoliMI SpringerBriefs, https://doi.org/10.1007/978-3-031-80268-3_10

opinion has already documented cases of altered satellite imagery [3]. Consequently, the field of Multimedia Forensics (MMF) has gained critical importance. MMF focuses on automatically assessing the integrity of multimedia objects, including images, audio clips, and videos. Researchers can reconstruct an object's lifecycle and expose manipulations by identifying traces, called Forensic Footprints (FFs), left by editing operations. However, data-driven techniques, particularly Neural Networks (NNs) and Convolutional Neural Networks (CNNs), now lead the way in forensic tasks, outperforming traditional signal processing methods [20].

Despite their resemblance with digital pictures, satellite images represent signals with a very different lifecycle. They are inherently a multimodal data asset due to the presence of various families of sensors onboard satellites. Moreover, creating a satellite image involves a series of complex signal-processing operations. Data modalities describing different information than standard natural photographs and more complicated processing pipelines imply that many forensic traces exploited in literature might be absent or not be performant enough to determine the authenticity of satellite images [1]. In this context, data-driven tools can be a valuable aid since they can automatically extract the features needed from corpora of data.

This chapter will show that CNNs are practical tools for extracting forensic features from satellite images. In particular, we will use CNNs to obtain some basic information about the input data, i.e., information about its source. We will show that we can extract such information at different granularities, from the satellite that generated it, to the product category to which a specific image belongs. We will then use it to determine the integrity of the object at hand, i.e., to localize splicing attacks, studying two different satellite imagery modalities: panchromatic and Synthetic Aperture Radar (SAR). Our techniques prove more reliable and performant than methods developed to analyze standard digital pictures. Moreover, to our knowledge, our work is the first documented in the literature on the forensic analysis of SAR images.

The material from this chapter summarizes part of the Ph.D. research presented in my thesis [4] and the following scientific papers [5, 7, 8]. We organize the rest of the chapter as follows. Section 2 provides a brief background on the different modalities of satellite imagery. Section 3 explains the basis of our forensic analyses, i.e., using CNNs to extract source attribution information from satellite images. Section 3.1 shows how we can use CNNs to attribute a panchromatic image to a specific satellite, while Sect. 3.2 illustrates a pipeline for extracting source-attribution features to localize splicing attacks. Section 3.3 analyzes the problem of splicing localization in SAR images. Finally, Sect. 4 concludes the chapter.

2   The Sophisticated World of Satellite Imagery

Remote sensing encompasses a variety of methods used to measure electromagnetic radiation as it interacts with the Earth's atmosphere and surface. There are two primary types of sensors used on satellites to collect remote sensing data: (i) active sensors, which have their own energy source to illuminate the object they are observing, and (ii) passive sensors, which depend on external sources like sunlight. Figure 1 provides a graphical explanation.

Electro-Optical (EO) imagery refers to satellite photos taken by passive sensors that detect sunlight reflected off the Earth's surface [1]. Passive sensors can capture wavelengths in the visible spectrum from Blue light, with a wavelength around 0,4 μm, up to Long-Wave Infrared wavelengths, i.e., 20 μm. These wavelengths are captured collectively or by using high-quality filters to separate them into different channels or bands. Practitioners consider imagery representing separate spectral content as distinct modalities. From this perspective, panchromatic is a modality of EO imagery capturing all the light in the visible spectrum, i.e., it is provided in a monochromatic format with no spectral information but high spatial resolution. RGB data collects information in the Red, Green, and Blue bands instead and is a valuable asset, together with other bands outside the visible spectrum, for land-cover discrimination [1]. Figure 2 provides some examples.

SAR imagery is an example of data generated through active sensors. A SAR system is a kind of radar that is installed on a moving platform, such as a satellite or aircraft. As it moves, the system emits high-power electromagnetic waves from its antenna. These waves interact with objects on the Earth's surface and return to the antenna with changes in amplitude and phase based on their properties, such as permittivity, geometry, and roughness [16]. The antenna gathers these reflected echoes, which are then processed to generate a SAR image. This image is a complex 2D matrix, typically displayed by examining the intensity values that approximate the reflectivity of the ground points. SAR imagery has gained popularity for various applications due to its ability to deliver high-resolution images irrespective of cloud cover, weather conditions, or daylight [16]. Please refer again to Fig. 2 for an example.

Both EO and SAR signals are not directly interpretable as they are but require complex processing chains to provide users with manageable data [16, 18]. According to the amount of processing executed on them, producers divide satellite images into products [16]. Each producer delivers products using proprietary algorithms and data formats [19]. This tendency starkly contrasts the generation of standard digital pictures, processed with simpler operations like demosaicing and color correction and usually available in industry-standardized formats. Moreover, it makes the world of satellite imagery varicolored, both in terms of data types and modalities, but also of FFs.

Fig. 1 Comparison between passive sensing (i.e., the Sun is the energy source of the reflected electromagnetic radiation) and active sensing (i.e., the satellite possesses a energy source)

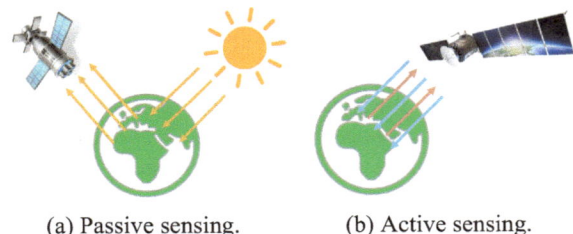

(a) Passive sensing.   (b) Active sensing.

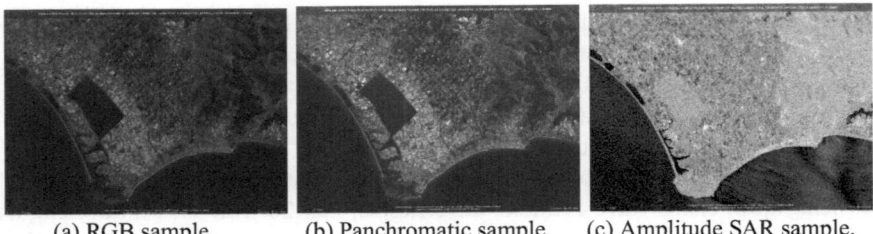

| (a) RGB sample. | (b) Panchromatic sample. | (c) Amplitude SAR sample. |

Fig. 2 Different modalities of satellite data from the same scene downloaded from the Copernicus Open Access Hub [10]. RGB and panchromatic are from the Landsat8 [19] mission. SAR is from Sentinel-1 [11]. We can already see some differences between the modalities, e.g., absence of spectral information in panchromatic, less spatial definition in RGB, speckle noise on radar signals, etc.

3 CNNs as Satellite Forensic Feature Extractors

CNNs have long been used in the forensic literature [20]. Their major appeal is that they ease the burden on practitioners to develop specific techniques for extracting forensic traces, i.e., CNNs learn the features needed for the task directly from data. Such property is particularly desirable given the complexity of satellite imagery generation pipelines. Therefore, in the following, we will show how CNNs are practical tools for extracting forensic information from satellite images. We will mainly focus on extracting image source attribution features. Image source attribution involves identifying the acquisition device used to capture the image under analysis. The literature has examined the problem at various levels for standard digital pictures [14], i.e., identifying the type of device used (e.g., camera vs. scanner), the specific brand or model (e.g., Sony vs. Canon), or the specific instance of the device (e.g., this iPhone X vs. that iPhone X). We can construct the same analogies for satellite images. We will first analyze how to attribute a panchromatic image to a specific sensor, i.e., a satellite, how we can use these features to highlight local manipulations, and finally how we can extract features specific to a particular amplitude SAR product to localize splicing attacks.

3.1 Panchromatic Imagery Open-Set Sensor Attribution

Our first study focuses on panchromatic imagery, addressing the issue of sensor attribution [5]. More specifically, we rely on CNNs to assign an image to the satellite that captured it. We consider two scenarios: closed-set attribution (where the sample originates from a known set of satellites) and open-set attribution (where if the image comes from an unknown satellite, the detector should be able to classify it as such). In the case of open-set attribution, we use ensemble techniques to gauge the confidence of our CNNs in performing the analysis. Specifically, we employ two model uncertainty estimation techniques: Deep Ensembles [2] and Monte Carlo Dropout [21]. Both strategies obtain multiple predictions for each image under analysis from an ensemble of CNNs. They then assess the consistency of the predictions as a measure of the ensemble's uncertainty. Intuitively, if all predictions concur (i.e., they all indicate the same satellite class), we trust the classification output (i.e., the image belongs to the satellite indicated by all networks). On the other hand, if the classification results are inconsistent, we conclude that none of the satellites has captured the image, and we attribute it to the unknown class (i.e., a satellite not in the training set). Figure 3 provides a graphical representation of the process. We tested this algorithm on a dataset of panchromatic images downloaded from various satellites. Our results highlight how this task is sensibly more complex than the corresponding device attribution for natural pictures, probably due to the more elaborated pipelines behind the generation of satellite images. Still, our technique obtains promising results and give us a robust descriptor for further analysis of the integrity of panchromatic samples, as we will show in the next Section.

3.2 Panchromatic Imagery Copy-Paste Localization

With "copy-paste", the forensic community refers to combining two pictures from different devices to create a "tampered" photo. Photoshopped images [17] are an example of such attacks, even though researchers might also refer to them as "splicing", i.e., attacks aimed at altering the content of a specific pixel region through data from the same or different samples and suitable editing functions.

Figure 4 provides a graphical representation of the splicing process. A region S from a source image $\mathbf{S}$ is selected and pasted on a region T of a target image $\mathbf{T}$. Some editing, such as resizing and rotation, might be applied to make the forgery more realistic. The final result is a tampered sample $\hat{\mathbf{T}}$, with a splicing mask $\mathbf{M}$

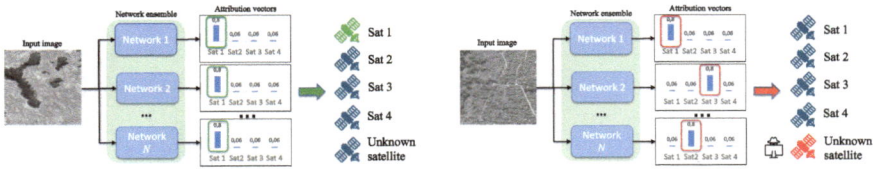

(a) Closed-set classification.   (b) Open-set classification.

Fig. 3 Visual representation of the proposed pipeline. Each CNN in the ensemble outputs an M−element vector that represents the likelihood for each satellite generating the input sample. Uncertainty is assessed based on the consistency of the scores. In Fig. 3a, all networks indicate satellite number 1, and the sample is thus classified. In Fig. 3b, the networks do not concur on the satellite that produced the image. As a result, the sample is rejected and allocated to the unknown class

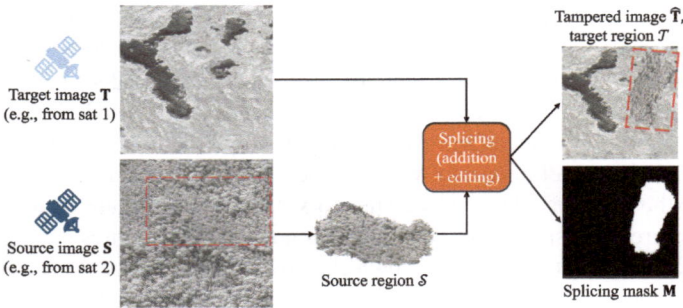

Fig. 4 Representation of the process behind a splicing attack

indicating the spliced pixels, i.e., where the image is compromised. In case **T** and **S** come from different sources, e.g., different camera models, satellites, etc., the attack is also referred to as copy-paste. The main goal of splicing localization is to localize which pixels of a tampered image $\hat{\mathbf{T}}$ are manipulated. Such a goal corresponds to estimating a tampering mask $\hat{\mathbf{M}}$ whose values are as close as possible to **M** from the sole analysis of $\hat{\mathbf{T}}$.

Building on the insights from Sect. 3.1, we developed a pipeline to identify copy-paste attacks in panchromatic images [8]. We train a group of CNNs, where each element extracts satellite-attribution features in a patch-wise fashion. We then combine this information to create a heatmap that pinpoints copy-paste attacks as spatial inconsistencies in the sensor-related data. Figure 5 provides a graphical representation of the full pipeline. We evaluated this approach against a dataset of panchromatic satellite copy-paste attacks. The results demonstrate that the proposed method surpasses more advanced image forensic tools designed for conventional photographs. Moreover, even though we do not train the CNNs with data-augmentation, our pipeline proves robust to various operations applied on the target region, e.g., blurring, contrast, resizing, etc. We believe such property derives from the complex generation pipeline of satellite data and that CNNs acquire it while extracting sensor-related features.

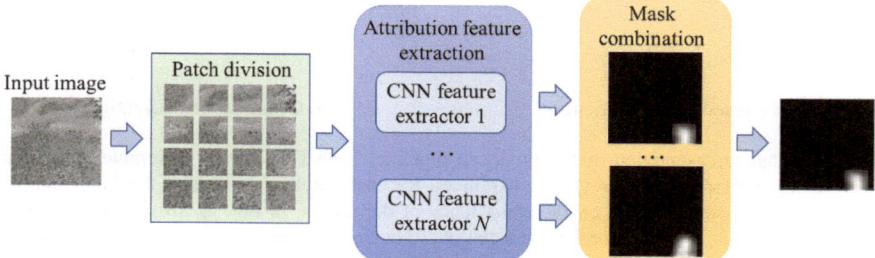

Fig. 5 Proposed panchromatic copy-paste localization pipeline. © IEEE 2022 [8]

3.3   CNN-Based Amplitude SAR Imagery Splicing Localization

As highlighted in Sect. 2, there are numerous differences between natural photographs and SAR products, the primary one being that SAR images are complex signals, meaning they convey both magnitude and phase information. However, it is also true that amplitude products, i.e., SAR images where only the magnitude is preserved, are widely used for various applications [16]. For an end-user who is not familiar with overhead imagery, amplitude SAR products can be viewed as natural photographs. Provided they have a single polarization, these products present amplitude information as a matrix of real numbers. As a result, any standard image editing tool can manipulate them to create spliced amplitude SAR images [7].

In [7], therefore, we tackle the problem of amplitude SAR imagery splicing localization. Our goal is to pinpoint which pixels are manipulated in an image tampered with a splicing attack, i.e., the insertion of a portion from a different source amplitude SAR image into a target SAR image. We also presume that image processing operations, such as blurring, resizing, noise addition, and others, might have modified the target region to make the attack more convincing and appealing.

We propose a CNN solution that extracts a fingerprint highlighting manipulation traces in the sample under analysis. The fingerprint exploits the concept of high-frequency residual, i.e., it is a heatmap generated by high-pass filtering the image under analysis through a De-noiser Convolutional Neural Network (DnCNN) [22]. The motivations behind this choice are twofold: (i) forensic traces become more noticeable when examining the high-frequency content of the picture; (ii) the amplitude SAR products we investigate undergo various operations from their capture to their final production [11, 16], such as resampling, de-ramping, ground-range projection, etc. As a result, we can infer that different products bear different processing traces. Given the nature of SAR signals and the non-linear operations employed, even amplitude SAR products from the same satellite may exhibit different traces relative to the processing performed to create them.

Consequently, we train the DnCNN to extract fingerprints that are as similar as possible only if the images in the training set originate from the same initial products. Whenever an attacker alters an image, editing traces corresponding to different acquisition or processing histories become visible in the fingerprint. We can further process the fingerprint to create a binary tampering mask highlighting the pixel region under attack. Figure 6 illustrates the complete pipeline. This work is the first contribution towards the forensic analysis of SAR imagery and outperforms state-of-the-art tools developed for conventional digital photographs on a dataset of spliced amplitude SAR images with different editing applied to the target region.

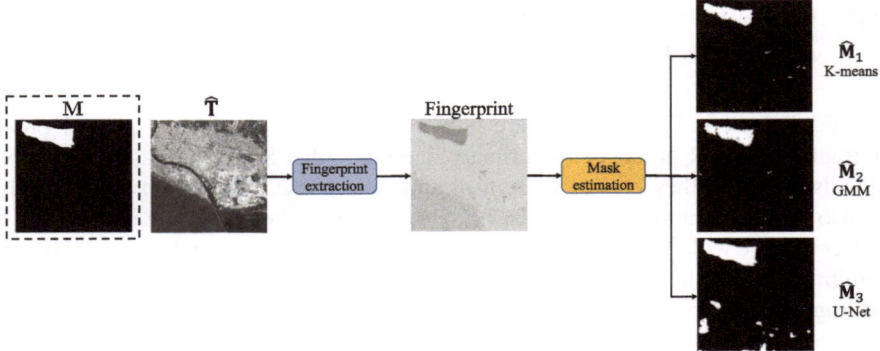

Fig. 6 Proposed amplitude SAR imagery splicing localization pipeline. We can already spot the tampered region $\mathcal{T}$ highlighted in the fingerprint due to the presence of different processing traces. Different methods after generate a mask estimate $\hat{\mathbf{M}}$

4 Conclusions

This chapter discussed the challenges faced by MMF practitioners when analyzing satellite imagery. The multimodal nature of this data asset, together with a complex generation pipeline, makes the FFs developed for standard digital pictures unsuitable for assessing its authenticity. However, CNNs are viable solutions to extract forensic information from different satellite imagery modalities. We showed how CNNs can extract features relative to the source of a specific sample, either the satellite generating it or a specific product the image has been derived from. We then applied such information to verify the integrity of panchromatic images, i.e., verifying if the input probe came from a set of known satellites and has been spliced, and amplitude SAR images, i.e., localizing splicing attacks. These works constitute the first in the literature regarding the analysis of such imagery. While presenting different processing pipelines, the evolution of the generation processes of multimedia objects forces MMF to maintain an open approach and seek similarities between distant data modalities. This is the case, for instance, of the traces left by super-resolution methods in satellite images and synthetic image generation techniques [6]. Therefore, future studies will be devoted to deepening the forensic knowledge of the generation pipelines of satellite images while contemporarily developing a fully multimodal approach following the recent trends in foundational models for remote sensing [13].

References

1. Abady, L., Cannas, E.D., Bestagini, P., Tondi, B., Tubaro, S., Barni, M.: An overview on the generation and detection of synthetic and manipulated satellite images. APSIPA Trans. Signal Inf. Process. **11**(1) (2022)
2. Balaji, L., Pritzel, A., Blundell, C.: Simple and scalable predictive uncertainty estimation using deep ensembles. In: International Conference on Neural Information Processing Systems (NeurIPS) (2017)
3. BBC: Conspiracy Files: Who shot down MH17?, April 2016. https://www.bbc.com/news/magazine-35706048. Accessed 11 Aug. 2023
4. Cannas, E.D.: Multimedia forensics challenges in the multimodality data era. Ph.D. thesis, Politecnico di Milano (2024)
5. Cannas, E.D., Baireddy, S., Bartusiak, E.R., Yarlagadda, S.K., Montserrat, D.M., Bestagini, P., Tubaro, S., Delp, E.J.: Open-set source attribution for panchromatic satellite imagery. In: International Conference on Image Processing (ICIP), pp. 3038–3042. IEEE (2021)
6. Cannas, E.D., Beaus, P., Bestagini, P., Marques, F., Tubaro, S.: A one-class approach to detect super-resolution satellite imagery with spectral features. In: International Conference on Acoustics, Speech and Signal Processing (ICASSP), pp. 4685–4689. IEEE (2024)
7. Cannas, E.D., Bonettini, N., Mandelli, S., Bestagini, P., Tubaro, S.: Amplitude SAR imagery splicing localization. IEEE Access **10**, 33882–33899 (2022)
8. Cannas, E.D., Horváth, J., Baireddy, S., Bestagini, P., Delp, E.J., Tubaro, S.: Panchromatic imagery copy-paste localization through data-driven sensor attribution. In: International Conference on Acoustics, Speech and Signal Processing (ICASSP), pp. 2889–2893. IEEE (2022)
9. CNN: New satellite images show buildup of Russian military around Ukraine, March 2022. https://edition.cnn.com/2022/02/02/europe/russia-troops-ukraine-buildup-satellite-images-intl/index.html. Accessed 11 July 2023
10. ESA: Copernicus Open Access Hub. https://scihub.copernicus.eu/. Accessed 2 Jan. 2022
11. ESA: Sentinel-1 Mission User Guide. https://sentinel.esa.int/web/sentinel/user-guides/sentinel-1-sar. Accessed 2 Jan. 2022
12. ESA: Monitoring Earths surface. https://www.esa.int/Enabling_Support/Preparing_for_the_Future/Discovery_and_Preparation/Monitoring_Earth_s_surface. Accessed 24 July 2022
13. Guo, X., Lao, J., Dang, B., Zhang, Y., Yu, L., Ru, L., Zhong, L., Huang, Z., Wu, K., Hu, D., He, H., Wang, J., Chen, J., Yang, M., Zhang, Y., Li, Y.: Skysense: a multi-modal remote sensing foundation model towards universal interpretation for earth observation imagery. In: IEEE/CVF Conference on Computer Vision and Pattern Recognition (CVPR), pp. 27672–27683 (2024)
14. Kirchner, M., Gloe, T.: Forensic camera model identification. In: Handbook of Digital Forensics of Multimedia Data and Devices. Wiley Ltd. (2015)
15. LandSat: U.S. geological survey—landsat data access, March 2022. https://www.usgs.gov/landsat-missions/landsat-data-access. Accessed 14 Mar. 2022
16. Oliver, C., Quegan, S.: Understanding Synthetic Aperture Radar Images. SciTech Publishing, SciTech radar and defense series (2004)
17. Paris, B., Donovan, J.: Deepfakes and cheap fakes. Data Soc. (2019)
18. Lillesand, T., Kiefer, R.W.: Remote Sensing and Image Interpretation. Wiley (2015)
19. U.S. Geological Survey: Landsat 8 data users handbook (2019). https://www.usgs.gov/media/files/landsat-8-data-users-handbook. Accessed 5 July 2023
20. Verdoliva, L.: Media forensics and deepfakes: an overview. J. Sel. Top. Signal Process. (2020)
21. Gal, Y., Ghahramani, Z.: Dropout as a Bayesian approximation: representing model uncertainty in deep learning. In: International Conference on Machine Learning (ICML) (2016)
22. Zhang, K., Zuo, W., Chen, Y., Meng, D., Zhang, L.: Beyond a Gaussian denoiser: residual learning of deep CNN for image denoising. IEEE Trans. on Image Process. **26**(7), 3142–3155 (2017)

Digital Signal Processing Methodologies for Audio System Modeling and Audio Output Enhancement

Riccardo Giampiccolo🄳

Abstract This chapter summarizes some of the main results I obtained during my Ph.D. studies at Politecnico di Milano, under the supervision of Professor Augusto Sarti and Professor Alberto Bernardini. Audio systems have become, nowadays, pervasive in many different market sectors, such as that of consumer electronics and biomedical devices. To accurately represent, digitally replicate, and process the signals of such complex systems, it is crucial to develop multiphysics models that capture their nonlinear behaviors. In this chapter, I introduce novel multiphysics models of audio systems together with innovative digital signal processing techniques with the ultimate goal of improving their acoustic response. By leveraging the efficiency and accuracy of Wave Digital Filters, I propose novel modeling methods to incorporate the various physical domains involved in audio systems. I introduce iterative methods for streamlined emulation and parallel implementation, addressing, at the same time, complex nonlinearities such as magnetic saturation and hysteresis. I present new linearization and virtualization algorithms to manipulate the audio device behavior leveraging on the newly introduced multiphysics models. Finally, I combine psychoacoustic methodologies with deep-learning models to tackle operating conditions affected by very strict physical constraints. With this investigation, I cover diverse audio signal processing tasks, offering fresh insights and practical solutions across various application scenarios.

Keywords Audio signal processing · Multiphysics models · Wave digital filters · Virtualization algorithms · Psychoacoustics · Deep learning

R. Giampiccolo (✉)
Dipartimento di Elettronica, Informazione e Bioingegneria, Politecnico di Milano, Via Ponzio 34/5, 20133 Milan, Italy
e-mail: riccardo.giampiccolo@polimi.it

S. Garatti (ed.), *Special Topics in Information Technology*,
PoliMI SpringerBriefs, https://doi.org/10.1007/978-3-031-80268-3_11

1   Introduction

The performance of audio systems, such as audio transducers and sensors, strictly depends on the nonlinearities affecting the involved physical domains, as well as on how those domains interact. For example, the weight of producing a desirable tone in audio amplifiers (e.g., vacuum tube guitar amplifiers) falls heavily on its transformers, whose magnetic characteristics typically exhibit hysteresis and saturation, but also on the nonlinear interaction between electrical and mechanical domains that often results in unpleasant distortions. The dynamics of multiphysics systems becomes even more complex when the size of the device is minimized. It is the case of small-size audio transducers, such as piezoelectric or MEMS (Micro-ElectroMechanical Systems) loudspeakers, in which companies, over the past few years, have been heavily investing. For instance, the number of consumer electronics devices integrating piezoelectric transducers as flat-panel loudspeakers has experienced a steep increase. In fact, being very thin, piezoelectric transducers cope with the miniaturization process that characterizes the market. However, at the same time, their reduced dimensions cause the sound pressure level to be poor at low frequency, impairing the overall acoustic response. In order to derive signal processing algorithms able to virtualize, compensate, or linearize the transducer behavior, multiphysics models must be derived. Moreover, such models should be as accurate as possible and, at the same time, characterized by a low computational cost, such that they can be used for real-time applications.

This chapter focuses thus on the multiphysics modeling of audio systems targeting mostly consumer electronics and Virtual Analog (VA) applications (i.e., the realization of digital counterparts of analog audio gear). I address such a task making use of Wave Digital Filters (WDFs), which are making headway among audio signal processing techniques thanks to their efficiency, stability, and accuracy. In the following, I provide some introductory arguments on the motivation behind the choice of WDFs for addressing physical modeling.

1.1   Overview on Circuit Simulation Methods

The physical modeling of audio systems can be addressed following both distributed [6] and lumped approaches [3]. The former rely on sets of Partial Differential Equations (PDEs) for deriving highly accurate models; the latter, instead, rely on sets of Ordinary Differential Equations (ODEs). Common methods for solving PDEs are the Finite-Difference Time Domain (FDTD) method [22], the Finite Element Method (FEM) [2], Digital Waveguide Modeling (DWG) [21], and Modal Synthesis [19]. In this chapter, however, I am interested in exploring the use of Lumped-Element Models (LEMs) since these, although being generally less accurate, allow us to obtain simpler representations. It follows that starting from LEMs, it is usually possible to obtain more efficient digital realizations of the reference physical systems with respect to implementations of distributed models. As a primary objective, in fact, I strive to derive lightweight physical models of audio systems able to be run on-the-fly with the highest accuracy possible, in such a way that they can be eventually integrated into online digital signal processing algorithms. Among others, I take into account those audio systems that have as input or output variables electrical signals. It has been demonstrated that LEMs do suit the task of modeling such audio systems, providing highly powerful representations in low-frequency scenarios, and enabling an almost straightforward coupling with analog systems. It is the case, for example, of the well-established Thiele-Small model [3] which, for its range of validity, has shown remarkable accuracy and descriptive properties [3, 4].

LEMs are typically described making use of equivalent electrical networks. Over the years, different methods have been proposed for the numerical solution of circuits. Among the most popular methods, we can mention *Modified Nodal Analysis (MNA)*, *State-Space Method*, *Port-Hamiltonian Method*, and *Wave Digital Filters (WDFs)*. In the following section, I provide the motivations that led me to prefer WDFs over other LEM techniques.

1.2 Brief Overview on WDFs

WDFs were introduced by A. Fettweis in the 70s for realizing digital filters able to mimic the behavior of reference analog circuits while retaining their characteristic features, such as passivity and losslessness [7]. This particular class of digital filters involves a change of variables, passing from port voltages and currents (Kirchhoff variables[1]) to incident and reflected waves (wave variables) with the introduction of a free parameter per port called *port resistance*. Then, elements and topological inter-connections are described in a separate fashion by means of input-output scattering relations. It follows that the discretization of the element time derivatives happens block-wise, leading to several advantages [7]. Moreover, thanks to the change of variables and a proper setting of port resistances, dynamic elements discretized by means of stable discretization methods are realized as explicit Wave Digital (WD) blocks. Interconnections among elements are, instead, modeled as N-port junctions described by $N \times N$ scattering matrices, and are typically called *junctions*. At the time of their introduction, WDFs were only able to digitally realize linear filters characterized by simple topologies (series and parallel junctions) in a fully explicit fashion, i.e., without the need of any iterative solver. In fact, it is possible to resolve Delay-Free Loops (DFLs), which are present whenever implicit instantaneous relations among incident and reflected waves are formed, by properly setting port resistances. In the theory of WDFs, when WD structures present no DFLs are said to be *computable* and elements or port of junctions that do not present DFLs are said to be *adapted*. With the passing of time, research interest has also been geared toward solu-

---

[1] In WDF theory, we call *Kirchhoff domain* the domain of voltages and currents, which, in turn, are referred to as *Kirchhoff variables*.

tions of other kinds of circuits rather than just linear passive networks. In particular, it has been demonstrated that Binary Connection Trees (BCTs) can be exploited for the solution of circuits containing a single one-port nonlinearity without the need for iterative solvers, a property that cannot be found in approaches based on Kirchhoff variables, e.g., MNA.

Recently, many other theoretical advancements have been carried out with the aim of extending the class of circuits realizable by means of WDFs. For example, different WD models of nonlinear elements have been proposed, such as diodes and vacuum tubes; moreover, methods for the implementation of arbitrary lossless topological junctions that make use of graph representations of electrical circuits, have also been proposed. Finally, efficient iterative methods have been introduced for the solution of circuits containing multiple one-port nonlinearities [1, 5, 20].

The research described in this chapter summarizes the results obtained in my Ph.D. thesis [8] and the scientific publications [9–18]. The remainder of the chapter is divided into three parts: **Part I** entitled *Circuital Modeling* gives insights into the multiphysics modeling of audio systems in the Wave Digital domain; **Part II** entitled *Simulation* deals with new methods that can be exploited for the efficient simulation of WD models with multiple nonlinearities; **Part III** entitled *Processing* explains how the proposed models and methods can be useful for processing the signals of audio systems with the aim of improving their acoustic response.

2   Part I: Circuital Modeling

In this section, I summarize my findings on multiphysics models of audio systems, namely piezoelectric transducers, audio transformers, and guitar pickups.

2.1   Modeling Small-Size Loudspeakers

In [15], I take into account the modeling of small-size loudspeakers, in particular of piezoelectric (PE) loudspeakers. In fact, the consumer electronics market is geared toward compact and ultra-thin devices, and PE transducers, thanks to their thin profile, well suit such a market trend. I derive, for the first time, models of piezoelectric audio transducers in the WD domain such that they can be integrated into digital signal processing algorithms for improving their acoustic response. In order to accomplish such a goal, all the presented models are explicit and can be run in real-time; this is achieved by building on the characteristics of WDFs, e.g., modularity. In particular, I present several lumped models obtained considering thickness-mode and quasi-static regime hypotheses. Thereafter, I address the case of distributed models (known under the name of Mason's models), which are able to describe the behavior of the transducer outside of the quasi-static regime. I first propose a linearized version of Mason's model that can be implemented according to the traditional WDF theory.

Then, I show, in the case in which the linearization hypothesis does not hold true, how it is possible to realize in the WD domain the frequency-dependent elements characterizing Mason's model, contrary to what can be done in standard circuit simulation software.

2.2 Modeling Electromagnetic Systems

As other audio devices, I address the modeling of different electromagnetic systems used in audio applications [10, 11, 14, 18]. I started by analyzing the interesting case of audio transformers. An accurate and efficient modeling of such devices, in fact, is desirable, since the sound of audio loudspeakers and amplifiers is mostly due to the nonlinearities affecting their magnetic core. The models of audio transformers are derived in the WD domain through a multiphysics approach. In particular, I address the modeling of electric and magnetic domains in a separate fashion enabling a fine control on the nonlinearities and geometry of the magnetic core. In order to couple the electric subcircuits to magnetic subcircuits, I present a nonreciprocal junction with memory, which I called "Magnetic/Electric Junction," and provide different WD realizations according to the method chosen to discretize the time derivatives. Moreover, I consider two types of nonlinearities: core saturation and core hysteresis. I propose to model the first in an explicit fashion employing Canonical Piecewise-Linear functions (CPWL) tuned on magnetization curves of datasheets. Notably, a CPWL function is a 1st order interpolator able to outline a curve in a global fashion starting from a certain number of points on its characteristics. Hysteresis requires, instead, a greater modeling capability since it is a highly-complex nonlinear phenomenon with memory. I propose to exploit the generalization capabilities of recent deep learning methods, such as Recurrent Neural Networks (RNNs). In particular, I train a Preisach RNN on hysteresis measurements directly in the WD domain. This constitutes the first attempt of merging WDFs with RNNs for modeling nonlinearities with memory. Finally, under certain assumptions, I show how it is possible to confine the hysteretic nonlinearity—and thus the neural network—into one single nonlinear element and exploit WDFs to emulate the resulting multiphysics structure in a fully explicit fashion. I then present a new nonlinear model of guitar pickup systems. The model is able to exploit measurements of the integral of the voltage over time, taken at the terminal of a magnetic pickup, and encompass it into a nonlinear element, which can be then employed for simulation. The overall model also comprises tone and volume controls present along the typical analog processing chain of electric guitars, as well as the high input impedance downstream. The nonlinearity is modeled by means of a smoothed version of CPWL functions, which allows us to obtain an accurate representation.

3 Part II: Simulation

In this section, I first outline a WD iterative method able to solve multiphysics models in a modular fashion, and then I introduce strategies for further enhancing its efficiency in terms of Real-Time Ratio (RTR).

3.1 Wave Digital Methods for Multiphysics Simulation

In [10, 11], I propose a generalized version of the Scattering Iterative Method, which I called Hierarchical Scattering Iterative Method (HSIM), able to solve WD structures composed of an arbitrary number of nonlinear elements and junctions, enhancing the modularity of the representation. Such a method comes in handy for the simulation of multiphysics circuits since it is desirable to maintain a separate description of the different physical domains. I apply such a method for the emulation of audio transformers, whose magnetic core nonlinearity cannot be modeled by means of one single nonlinear element. I then propose to exploit parallel computing for enhancing the performance of HSIM [17]. In fact, such an iterative scheme is characterized by a higher number of *embarrassingly parallelizable* operations, i.e., operations completely independent one of the others, being thus able to be addressed by different workers. I analyze the computational flow of the method and I show that it can be parallelized following three main different strategies. By applying simple concepts borrowed from parallel computing theory, I prove that it is possible to increase the performance of the algorithm in terms of Real-Time Ratio (RTR); this constitutes a first in-depth analysis of parallel computing techniques applied in the WD domain. The presented methods can be exploited for the emulation of multiphysics circuits and can be integrated into the signal processing algorithms summarized in Sect. 4.

4 Part III: Processing

In this section, I introduce algorithms that exploit multiphysics models of audio transducers (both sensors and actuators) for deriving digital signal processing algorithms able to improve their acoustic response. Moreover, I present innovative psychoacoustics-based algorithms able to accomplish the same task when the physics of the transducer is too constrictive (e.g., in the case of small-size transducers).

4.1 Virtualization of Audio Transducers

In [12], I revise Leuciuc's theory on circuit inversion providing guidelines specific to the case of audio transducers. In particular, I take into account all the possible combinations of input and output variables, and, for each of them, I outline precise rules for deriving the inverse circuital model. In particular, the theory is based on a theoretical two-port element known in circuit theory as *nullor*. By augmenting the circuit with a nullor, it is possible to exploit the direct system to obtain its inverse. Different methods for implementing nullors are available in the literature on WDFs. Here, I propose a novel approach based on a double digraph decomposition of the connection network that is able to improve the efficiency of the WD method employed for the solution of the WD structure [9]. Then, drawing on the promising result obtained in [3, 4], I describe two processing chains for addressing the task of transducer virtualization—one for sensor virtualization, the other for actuator virtualization [12]. Such processing chains exploit the cascade of direct and inverse systems to cancel out the behavior of the transducer and apply the target behavior. In fact, I define the task of virtualization as the task of making a sensor/actuator sound as a target sensor/actuator. Finally, I apply the proposed algorithms to address the cases of compression driver linearization [12], capacitive microphone virtualization [12], and guitar pickup virtualization [14].

4.2 Virtual Bass Enhancement

In [13, 16], I introduce innovative techniques that make use of psychoacoustic effects for enhancing the perception of bass tones. In fact, due to physical limitations, small-size loudspeakers are not able to reproduce low frequencies with a suitable Sound Pressure Level (SPL) and virtualization algorithms may not be sufficient for compensating such misbehavior. I thus take into account Virtual Bass Enhancement (VBE) algorithms, which usually exploit the "missing fundamental" phenomenon for tricking the human brain into perceiving such missing frequencies. As a matter of fact, a pitch is not only perceived if present into the audio track but also thanks to the periodicity of its higher harmonics. Among the different methodologies, I take into account time-domain techniques since I aim at developing lightweight algorithms suitable to be integrated into the processing chains considered for consumer electronics applications. Time-domain techniques extract the low end of the audio track by means of a crossover network and achieve harmonic generation by processing it with nonlinear devices (NLDs). I propose a circuit equivalent representation of a typical generic time-domain VBE system opening up possibilities for designing new systems in both analog and digital domains [13]. Moreover, I introduce normalization stages for adapting the support of the nonlinear functions characterizing NLDs and better controlling the harmonic generation. Finally, I propose an innovative VBE system employing deep learning techniques for accomplishing source separation [16]. I substitute the typical crossover network with a neural network able to split the music track into several stems (e.g., bass, drums, vocals, etc.). I then describe a new processing pipeline that allows us to apply different types of NLDs and normalization functions to each of the obtained stems, sorting out most of the issues of time-domain techniques. Said VBE system turns out to outperform the state-of-the-art as far as both bass enhancement and audio quality are concerned.

5 Conclusions

This chapter summarized three years of doctoral studies conducted on the development of lumped multiphysics models of audio systems for audio output enhancement. Building on Wave Digital Filter (WDF) theory, I derived various models of audio systems including audio transformers, guitar pickups, and piezoelectric loudspeakers. These models were presented together with their nonlinearities, and iterative methods were developed for their efficient emulation, even taking advantage of modern multi-core DSPs. The models served as DSP modules in innovative (virtualization) algorithms designed for enhancing the acoustic performance of said audio systems. Moreover, I introduced new processing pipelines based on VBE systems to achieve the desired behaviors despite physical system limitations.

Many are the directions for further research, including the development of more refined WD multiphysics models for piezoelectric loudspeakers and the modeling of MEMS loudspeakers. Techniques for accommodating multiple nonlinear multi-port WD blocks in complex models and deriving parallel implementations with advanced scheduling policies are also envisioned. Additionally, other key areas for future research may concern extending the inversion theory to Multiple-Input Multiple-Output (MIMO) systems for processing arrays of audio sensors and actuators, developing new inversion chains for array processing, and designing novel virtual bass enhancement pipelines using deep learning models.

Overall, my doctoral research provides a robust foundation for advancing physics-based digital signal processing algorithms, with significant implications for enhancing the acoustic performance of audio systems across various applications, laying the groundwork for future innovations in audio technology.

References

1. Albertini, D., Bernardini, A., Sarti, A.: Scattering iterative method based on generalized wave variables for the implementation of audio circuits with multiple one-port nonlinearities. In: Proceedings of the 150th Audio Engineering Society Convention (2021)
2. Bathe, K., Wilson, E.: Numerical Methods in Finite Element Analysis. Prentice Hall, Englewood Cliffs (1976)

3. Bernardini, A., Bianchi, L., Sarti, A.: Loudspeaker virtualization–Part I: digital modeling and implementation of the nonlinear transducer equivalent circuit. Signal Process. **202**, 108720 (2022). https://doi.org/10.1016/J.SIGPRO.2022.108720
4. Bernardini, A., Bianchi, L., Sarti, A.: Loudspeaker virtualization–Part II: The inverse transducer model and the direct-inverse-direct chain. Signal Process. 108713 (2022). https://doi.org/10.1016/J.SIGPRO.2022.108713
5. Bernardini, A., Bozzo, E., Fontana, F., Sarti, A.: A wave digital newton-raphson method for virtual analog modeling of audio circuits with multiple one-port nonlinearities. IEEE/ACM Trans. Audio Speech Lang. Process. **29**, 2162–2173 (2021). https://doi.org/10.1109/TASLP.2021.3084337
6. Bilbao, S., Desvages, C., Ducceschi, M., Hamilton, B., Harrison-Harsley, R., Torin, A., Webb, C.: Physical modeling, algorithms, and sound synthesis: the NESS project. Comput. Music J. **43** (2020)
7. Fettweis, A.: Wave digital filters: theory and practice. Proc. IEEE **74**, 270–327 (1986). https://doi.org/10.1109/PROC.1986.13458
8. Giampiccolo, R.: Multiphysics modeling of audio systems in the wave digital domain. Ph.D. thesis, Politecnico di Milano, Ph.D. Programme in Information Technology (2023)
9. Giampiccolo, R., de Bari, M.G., Bernardini, A., Sarti, A.: Wave digital modeling and implementation of nonlinear audio circuits with nullors. IEEE/ACM Trans. Audio Speech Lang. Process. **29**, 3267–3279 (2021). https://doi.org/10.1109/TASLP.2021.3120627
10. Giampiccolo, R., Bernardini, A., Gruosso, G., Maffezzoni, P., Sarti, A.: Multiphysics modeling of audio circuits with nonlinear transformers. J. Audio Eng. Soc. **69**(6), 378–388 (2021). https://doi.org/10.17743/JAES.2021.0008
11. Giampiccolo, R., Bernardini, A., Gruosso, G., Maffezzoni, P., Sarti, A.: Multidomain modeling of nonlinear electromagnetic circuits using wave digital filters. Int. J. Circ. Theory Appl. **50**(2), 539–561 (2022). https://doi.org/10.1002/cta.3146
12. Giampiccolo, R., Bernardini, A., Massi, O., Sarti, A.: On the virtualization of audio transducers. Sensors **23**(11), 5258 (2023). https://doi.org/10.3390/s23115258, https://www.mdpi.com/1424-8220/23/11/5258
13. Giampiccolo, R., Bernardini, A., Sarti, A.: A time-domain virtual bass enhancement circuital model for real-time music applications. In: Proceedings of the 24th IEEE International Workshop on Multimedia Signal Processing (MMSP), pp. 1–5. Shanghai, China (2022). https://doi.org/10.1109/MMSP55362.2022.9949443
14. Giampiccolo, R., Bernardini, A., Sarti, A.: Virtualization of guitar pickups through circuit inversion. IEEE Signal Process. Lett. **30**, 458–462 (2023). https://doi.org/10.1109/LSP.2023.3269000, https://ieeexplore.ieee.org/document/10105937
15. Giampiccolo, R., Bernardini, A., Sarti, A.: Wave digital models of piezoelectric transducers for audio applications. IEEE Sens. J. **23**(1), 389–400 (2023). https://doi.org/10.1109/JSEN.2022.3225507
16. Giampiccolo, R., Mezza, A.I., Bernardini, A., Sarti, A.: Virtual bass enhancement via music demixing. IEEE Signal Process. Lett. **30**, 908–912 (2023). https://doi.org/10.1109/LSP.2023.3296877, https://ieeexplore.ieee.org/document/10187677
17. Giampiccolo, R., Natoli, A., Bernardini, A., Sarti, A.: Parallel wave digital filter implementations of audio circuits with multiple nonlinearities. J. Audio Eng. Soc. **70**(6), 469–484 (2022). https://doi.org/10.17743/jaes.2022.0012
18. Massi, O., Mezza, A.I., Giampiccolo, R., Bernardini, A.: Deep learning-based wave digital modeling of rate-dependent hysteretic nonlinearities for virtual analog applications. EURASIP J. Audio Speech Music Process. **2023**(1), 12 (2023). https://doi.org/10.1186/s13636-023-00277-8
19. Morrison, J.D., Adrien, J.M.: MOSAIC: a framework for modal synthesis. Comput. Music J. **17** (1993). https://doi.org/10.2307/3680569
20. Olsen, M.J., Werner, K.J., Smith, J.O.: Resolving grouped nonlinearities in wave digital filters using iterative techniques. In: Proceedings of the 19th International Conference on Digital Audio Effects (DAFx16) (2016)

21. Smith, J.O.: Physical modeling using digital waveguides. Comput. Music J. **16**, 74 (1992). https://doi.org/10.2307/3680470
22. Yee, K.S.: Numerical solution of initial boundary value problems involving Maxwell's equations in isotropic media. IEEE Trans. Antennas Propag. **14** (1966). https://doi.org/10.1109/TAP.1966.1138693

Optimized ISAC Waveform Design in 6G Networks

Silvia Mura

Abstract Sixth-generation (6G) wireless networks introduce Integrated Sensing and Communication (ISAC) technology, enabling simultaneous communication and sensing through shared time, frequency, space, and energy resources. Designing ISAC waveforms is challenging due to the need to satisfy both high-capacity communication and accurate sensing. This work presents a dual-domain waveform design that integrates classical Orthogonal Frequency Division Multiplexing (OFDM) with a tailored sensing signal in the delay-Doppler domain with carefully adjusted power to enhance sensing resolution without significantly affecting communication rate. Moreover, the paper tackles practical challenges such as high sidelobes and decreased sensing accuracy due to underutilized frequency-time resources. To address these issues, optimal resource allocation over time, frequency, and energy is defined, along with a novel interpolation technique based on Schatten p quasi-norm matrix completion. Numerical results show that these approaches outperform current methods, demonstrating the potential for improving 6G networks.

1 Introduction

Sixth-generation (6G) wireless networks are set to revolutionize the integration of radar sensing as a service, facilitated by utilizing millimeter-wave (mmWave) and massive antenna arrays [1]. Traditionally, radar systems use a dedicated spectrum to avoid interfering with communication. However, integrating radar within 6G networks requires a more efficient approach, as incorporating standalone radars within the communication infrastructure would be impractical and wasteful.

In response to this challenge, Integrated Sensing and Communication (ISAC) systems have emerged as a promising solution by using a single waveform for both communication and sensing over shared resources. However, designing ISAC waveforms is challenging due to differing performance requirements: communication systems prioritize reliable, high-capacity data transfer, while radar systems focus on

S. Mura (✉)
Politecnico di Milano, Piazza Leonardo da Vinci 32, 20133 Milan, Italy
e-mail: silvia.mura@polimi.it

© The Author(s) 2025
S. Garatti (ed.), *Special Topics in Information Technology*,
PoliMI SpringerBriefs, https://doi.org/10.1007/978-3-031-80268-3_12

target detection and localization. Therefore, Orthogonal Frequency Division Multiplexing (OFDM), the standard communication waveform, requires adjustments to support both communication and sensing functionalities effectively [9]. Recently, Orthogonal Time Frequency Space (OTFS) modulation has been explored for ISAC to address doubly-selective channels. However, its integration into 3GPP standards is difficult due to burst processing, which conflicts with 6G's low latency requirements [8].

Recent literature on ISAC waveform design spans multiple domains including space, frequency, and time, aiming to optimize the trade-off between communication capacity and sensing accuracy [4]. Information-theoretical approaches have provided useful bounds on the communication rate-sensing Cramér-Rao Bound (CRB) trade-off curve [10]. Spatial designs optimize antenna beam patterns for communication and sensing tasks [5], while time and frequency domain efforts, including dual delay-Doppler (DD) domain designs, enhance OFDM waveforms to improve sensing capabilities while maintaining compatibility with existing standards. Strategies such as subcarrier allocation, power optimization, and ambiguity function control are explored to achieve efficient resource utilization and robust performance [4].

However, waveform design approach performances vary significantly when resources are underutilized (i.e., resource occupancy factor (ROF) $\leq 100\%$). Low ROF levels, common in practical network scenarios, cause significant sidelobes in the ambiguity function, which impact sensing accuracy. Recent efforts address this challenge by proposing techniques to mitigate sidelobes and optimize sensing performance under limited resource conditions [1, 7]. This paper introduces a novel ISAC waveform by superimposing conventional OFDM with a dedicated sensing signal. Optimized for power, this sensing signal enhances ISAC system resolution while preserving communication rates and maintaining OFDM compatibility. Additionally, an ISAC waveform design for OFDM systems is proposed, optimizing time-frequency and energy to minimize delay and Doppler CRBs while ensuring communication rate QoS within limited ROF constraints. Sidelobe levels are managed using a Schatten p-norm sensing channel interpolation technique. These approaches significantly improve delay and Doppler CRBs and detection performance compared to conventional methods under low ROF conditions.

2   System Model

The ISAC system in Fig. 1 is considered, where the base station (BS) serves K single antenna users' equipment (UEs) while simultaneously sensing the environment. For simplicity, UEs are the sole targets, and spatial precoding/decoding is not considered, although it can be easily included. The ISAC BS employs an OFDM waveform with time and frequency resources partitioned into M subcarriers, spaced by Δf, and N OFDM symbols/time slots with duration $T = 1/\Delta f$. The overall bandwidth is

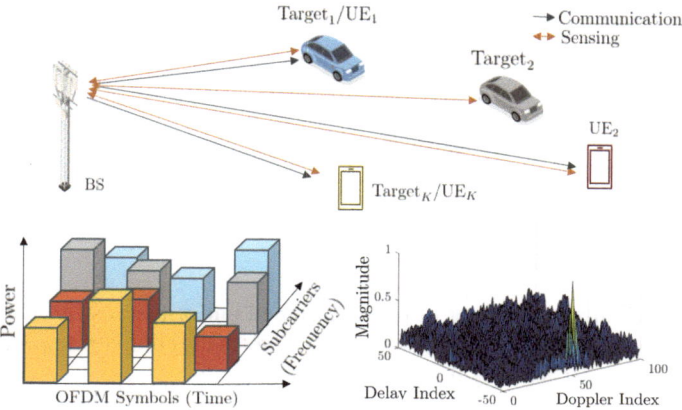

Fig. 1 The standard OFDM waveform with low ROF shows high sidelobes in its 2D ambiguity function, negatively affecting sensing performance in a multi-target/UE scenario

$B = M\Delta f$, and the ISAC burst duration is NT. The transmitting (Tx) signal in frequency-time (FT) domain is represented by $\mathbf{X} \in \mathbb{C}^{M \times N}$, whose (m, n)th element is

$$[\mathbf{X}]_{m,n} = [\mathbf{\Sigma} \odot \mathbf{S}]_{m,n} = \sigma_{m,n} s_{m,n}, \tag{1}$$

where $[\mathbf{\Sigma}]_{m,n} = \sigma_{m,n} \in \mathbb{R}_+$ is the square root of the allocated energy to the (m, n)th communication symbol $[\mathbf{S}]_{m,n} \in \mathbb{C}$. The total energy of the Tx signal is $E_{tot} = \|\mathbf{X}\|_F^2 = \|\mathbf{\Sigma}\|_F^2$. The received sensing signal matrix $\mathbf{R} \in \mathbb{C}^{M \times N}$ at the BS in the FT domain is

$$\mathbf{R} = \mathbf{X} \odot \mathbf{H}_s + \mathbf{W}, \tag{2}$$

where $\mathbf{H}_s \in \mathbb{C}^{M \times N}$ represents the sensing channel capturing the echos from the K UEs/targets and $\mathbf{W} \in \mathbb{C}^{M \times N}$ gathers the noise samples, statistically uncorrelated across time and frequency. The sensing channel $\mathbf{H}_s$ pertaining to the (m, n)th resource is

$$[\mathbf{H}_s]_{m,n} = \sum_{k=1}^{K} \beta_k \, e^{j2\pi(\nu_k nT - \tau_k m\Delta f)}, \tag{3}$$

where (i) $\beta_k \sim \mathcal{CN}(0, \Omega_\beta^{(k)})$ represents the complex scattering amplitude for the kth UE/target, with $\Omega_\beta^{(k)}$ proportional to $f_0^{-2} R_k^{-4}$ and dependent on the carrier frequency f_0, the distance R_k between the BS and the kth UE/target, and the target's reflectivity; (ii) $\tau_k = 2R_k/c$ is the propagation delay for the kth UE/target; (iii) $\nu_k = 2f_0 V_k/c$ denotes the Doppler shift due to the radial velocity V_k of the kth UE/target. The maximum delay, $\tau_{max} = \max_k(\tau_k)$, is constrained to be less than the cyclic prefix duration T_{cp}, ensuring $\tau_{max} \leq T_{cp}$ for unambiguous range estimation.

The FT communication received signal at the kth UE within the (m, n)th resource bin is

$$\mathbf{Y}_k = \mathbf{X} \odot \mathbf{H}_k + \mathbf{Z}, \tag{4}$$

and the communication channel is defined as

$$[\mathbf{H}_k]_{m,n} = \sum_{q=1}^{Q} \alpha_q^{(k)} \, e^{j2\pi(v_q^{(k)} nT - m\Delta f \tau_q^{(k)})}, \tag{5}$$

where Q represents the number of paths, assumed uniform across all UEs for simplicity. For each qth path, $\alpha_q^{(k)} \sim \mathcal{CN}(0, \Omega_q^{(k)})$ denotes the complex amplitude for the kth UE. $\tau_q^{(k)}$ and $v_q^{(k)}$ indicate the delay and Doppler shift, respectively, for the qth path of the kth UE. Unlike the sensing signal $\mathbf{R}$ at the BS, the communication channel does not preserve the actual delay and Doppler shifts due to FT synchronization by the UE terminal. The additive noise, $z_{m,n}^{(k)}$, is uncorrelated across UEs, time, and frequency.

3 Dual-Domain Waveform Superposition

This section discusses the fundamentals of dual-domain waveform design, as depicted in Fig. 2. The Tx ISAC signal matrix in the FT domain is constructed through the following superposition:

$$\mathbf{X} = \mathbf{\Sigma}_{\text{com}} \odot \mathbf{S}_{\text{com}} + \sigma_{\text{sen}} \mathbf{S}_{\text{sen}}, \tag{6}$$

where: (i) $\mathbf{S}_{\text{com}} \in \mathbb{C}^{M \times N}$ is the communication signal matrix for all K UEs, (ii) $\mathbf{S}_{\text{sen}} \in \mathbb{C}^{M \times N}$ is the sensing signal matrix, (iii) $\mathbf{\Sigma}_{\text{com}}$ represents the square root of the allocated power for the communication signal, and (iv) $\sigma_{\text{sen}} \geq 0$ is the amplitude of the sensing signal. The OFDM communication signal $\mathbf{S}_{\text{com}}$ exhibits a ROF of 20–50% [2]. The sensing signal is designed as a single pulse with unit amplitude in the DD domain as

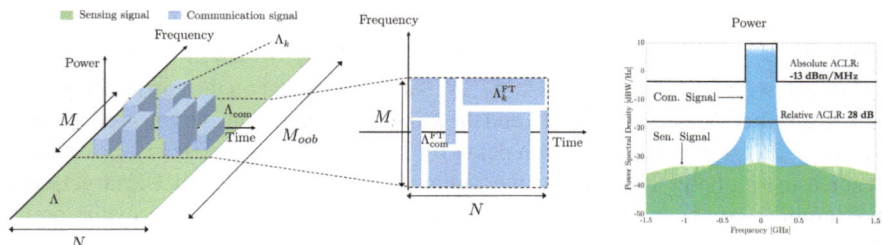

Fig. 2 FT-power resource allocation representation for the dual-domain waveform design

$$\left[\tilde{\mathbf{S}}_{\text{sen}}\right]_{\ell,p} = \delta_{\ell-\ell_i,\,p-p_i}, \tag{7}$$

(ℓ_i, p_i) represents the position of the Tx sensing pulse in the DD domain. The BS maps this pulse to the FT domain using a DFT-IDFT pair along the delay and Doppler dimensions:

$$\mathbf{S}_{\text{sen}} = \mathbf{F}_M \tilde{\mathbf{S}}_{\text{sen}} \mathbf{F}_N^{\text{H}}, \tag{8}$$

where $\mathbf{F}_M \in \mathbb{C}^{M \times M}$ and $\mathbf{F}_N \in \mathbb{C}^{N \times N}$ are DFT matrices and the (m, n)th entry of $\mathbf{S}_{\text{sen}}$ is the sample of a 2D sinusoid.

Enhancing range resolution requires surpassing regulatory bandwidth limits for the sensing signal. However, it must comply with bandwidth-integrated power constraints specified by the adjacent channel leakage ratio (ACLR) [2], defined as

$$\frac{P_{\text{ib}}}{P_{\text{oob}}} \geq \text{ACLR}_{\text{rel}}, \quad (\sigma_{\text{sen}})^2 \leq \text{ACLR}_{\text{abs}} \Delta f. \tag{9}$$

The first constraint limits the ratio of in-band to out-of-band (IB-to-OOB) power. In-band power includes the sum of communication and sensing signals within the legacy bandwidth, while out-of-band power pertains to the sensing signal beyond this range, considering the total bandwidth $M_{oob}\Delta f$, as shown in Fig. 2 ($\text{ACLR}_{\text{rel}} \in [22, 28]$ dB [2]). The second constraint caps the absolute power spectral density of the OOB signal at $\text{ACLR}_{\text{abs}} = -13$ dBm/MHz. Thus, the achievable range resolution with OOB emission becomes $\Delta R = c/(2M_{oob}\Delta f) > c/(2M\Delta f)$, which is an improvement compared to using the communication signal only.

4 Optimized OFDM ISAC Waveform Design with Limited Resource Occupancy

This section introduces an OFDM-optimized waveform without superposition, focusing on joint frequency-time-energy allocation to minimize CRBs for delay and Doppler ($\mathbf{C}\tau$ and $\mathbf{C}\nu$), defined as in [4]. The optimization problem involves allocating energy $\mathbf{e}_k \in \mathbb{R}^{L \times 1}$ per resource to the kth UE and assigning Boolean FT allocation variable $\mathbf{a}_k \in \mathbb{B}^{L \times 1}$, with $L = m, n$. The optimization is formulated as:

$$\min_{\{\mathbf{e}_k,\mathbf{a}_k\}_{k=1}^K} \quad \epsilon_\tau \, \mathrm{tr}\left(\frac{\mathbf{C}_\tau}{\Delta\tau^2}\right) + \epsilon_\nu \mathrm{tr}\left(\frac{\mathbf{C}_\nu}{\Delta\nu^2}\right)$$

$$\text{s. t.} \quad \frac{1}{L}\sum_{\ell=1}^L \log_2(1 + \gamma_{k,\ell}) \geq \bar{\eta}, \quad \sum_{k=1}^K [\mathbf{a}_k]_\ell \leq 1,$$

$$\sum_{k=1}^K \mathbf{1}^T \mathbf{a}_k \leq \mu L, \quad \sum_{k=1}^K \mathbf{1}^T \mathbf{e}_k \leq E_{\text{tot}},$$

$$0 \leq [\mathbf{e}_k]_\ell \leq [\mathbf{a}_k]_\ell \, \sigma_{\max}^2, \quad \left|[\mathbf{e}_k]_\ell - [\mathbf{e}_k]_{\ell+1}\right| \leq \Delta T,$$

$$\forall \ell, \forall k, \tag{10}$$

where ϵ_τ and ϵ_ν denote the weights of the CRBs, and Δ_τ and Δ_ν indicate the corresponding resolutions. The defined optimization problem guarantees the achievable spectral efficiency QoS $\bar{\eta}$, where the Signal-to-Noise ratio (SNR) is $\gamma_{k,\ell} = [\mathbf{e}_k \odot \mathbf{a}_k]_\ell |H_k|^2/N_0$ and $|Hk|^2 = \mathbb{E}_\alpha\left[\|\mathbf{H}_k\|_F^2\right]/L$. Moreover, the constraints of the optimization problem ensure interference avoidance, adherence to a specified ROF level (μ), compliance with the total energy budget, and a maximum energy limit per resource σ_{max}^2. Finally, smooth energy transitions are enforced to minimize sidelobes and ensure gradual energy changes (ΔT), which impact sensing accuracy and communication quality. Although the problem is non-convex, it can be relaxed into a convex form, as outlined in [6], and formulated as a Mixed Integer Cone Program solvable by Branch and Bound technique.

Figure 3 displays the waveforms designed across the energy, time, and frequency domains, with resource allocation at grid boundaries to reduce CRBs. It also illustrates the influence of the maximum energy gradient Δ on CRBs, highlighting the strategic allocation of higher power levels at the grid edges to further minimize CRBs. However, holes in the waveform, as shown in Fig. 3a, cause high sidelobes that affect target resolution. To address this, a sensing channel interpolation technique using Schatten p-norm matrix completion is proposed

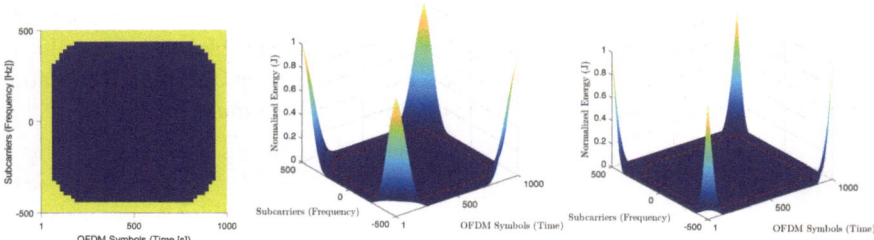

Fig. 3 Optimized waveform for $\mu = 0.25$: from left to right, **a** FT resource allocation (blue: unused, yellow: allocated), **b** frequency-time-energy allocation with $\Delta = -30$ dB, **c** frequency-time-energy allocation with $\Delta = -10$ dB

Table 1 Simulation parameters

Parameter	Symbol	Value (s)
Carrier frequency	f_0	30 GHz
Subcarrier spacing	Δf	1 MHz
Number of subcarriers/OFDM symbols	M, N	1000
OFDM occupancy ratio	μ	0.1–0.5
Range	R	10–60 m
Total energy	E_{tot}	$43 \times T$ dBmJ

$$\min_{\mathbf{H}_s} \ ||\mathbf{H}_s||_p^p \text{ s.t. } ||[\mathbf{H}_s]_\mathbf{A} - [\widehat{\mathbf{H}}_s]_\mathbf{A}||_2 \le \epsilon,$$

where $|| \cdot ||_p^p$ denotes the Schatten p-norm as in [6], ϵ is a small positive constant, and $[\widehat{\mathbf{H}}s]_\mathbf{A}$ represents the estimated channel samples from the least squares approach, with resource allocation $\mathbf{A} = \text{vec}^{-1}(\sum_k \mathbf{a}_k)$. The Schatten p-norm balances convexity and rank approximation, with $p = 1$ yielding a convex problem and $p \to 0$ providing a more accurate matrix rank estimate.

5 Numerical Results

This section evaluates the proposed ISAC waveform by comparing the dual-domain waveform with conventional OFDM and OTFS [8]. It also assesses the optimized OFDM ISAC waveform with limited resource occupancy against standard random and contiguous resource scheduling [3], all under the same ROF μ and constant energy per resource. Simulation parameters are detailed in Table 1. The numerical results include the sensing CRB gain, defined as the ratio of the benchmarks' root CRB on delay to that of the proposed waveform, with values >1 in linear scale or 0 dB indicating improvement. Communication performance is evaluated by the achievable rate η (bits/s/Hz). Finally, the ambiguity function performance is reported to show sidelobe level reduction.

Figure 4a shows the CRB gain of the dual-domain waveform relative to benchmarks (OFDM [3] and OTFS [8]) based on ROF μ and fractional communication bandwidth M/M_{oob}. For M/M_{oob} $\le$20–30%, the dual-domain ISAC design improves CRB over OFDM and OTFS, while at higher bandwidths, OFDM and OTFS are preferable. Figure 4 (b) shows that the achievable rates of the dual-domain waveform, OFDM, and OTFS systems vary with transmission power and distance. The dual-domain waveform incurs a minimal penalty compared to OFDM, noticeable only at $\eta \ge 8$ bps/Hz and decreasing with higher BS-UE distance. OTFS excels at medium to high achievable rates due to its lack of a cyclic prefix. The superiority of

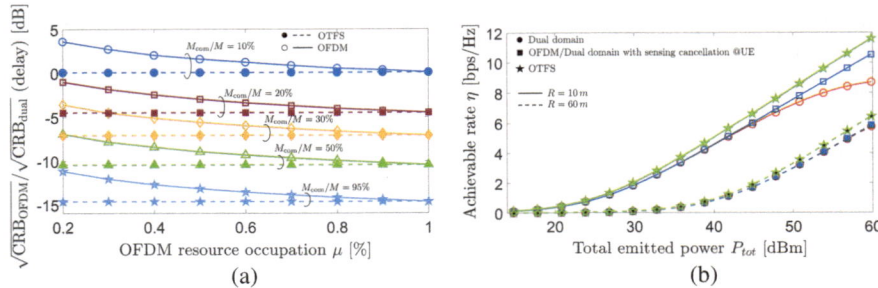

Fig. 4 Dual-Domain waveform performance: (from left to right). **a** CRB gain, **b** achievable rate

the dual-domain waveform is evident in Fig. 5, which shows reduced sidelobe levels and a sharper main lobe compared to OFDM.

Figure 6a shows the CRB gain for the optimized OFDM ISAC waveform with varying ROF, plotted on a linear scale, against inter-delay spacing normalized to the delay resolution $\Delta\tau$. With a bandwidth occupancy factor $\mu = 0.25$, the proposed ISAC waveform outperforms standard random and contiguous scheduling, as

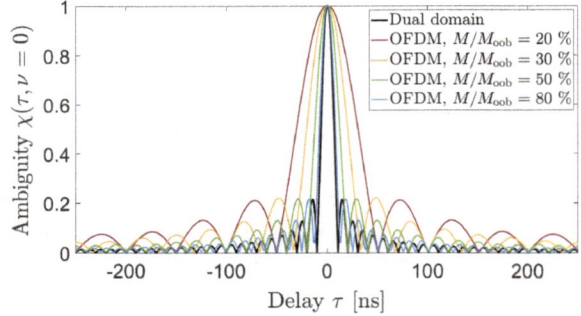

Fig. 5 Ambiguity function for dual-domain and OFDM with different M/M_{oob}

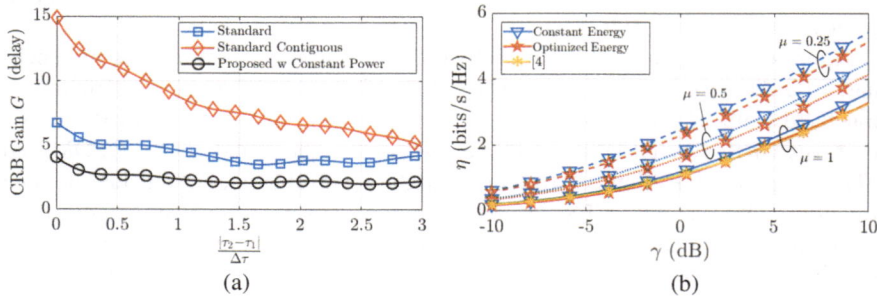

Fig. 6 Optimized waveform performance: (from left to right). **a** CRB Gain, **b** achievable rate

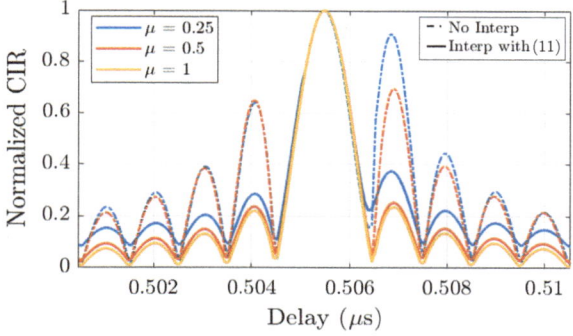

Fig. 7 CIR before and after interpolation for different ROF with respect to $\mu = 1$

well as a constant power waveform optimized only for time and frequency, achieving CRB gains of $6\times$, $14\times$, and $5\times$, respectively, for closely spaced targets. This demonstrates its effectiveness in distinguishing closely placed targets. Additionally, optimizing energy allocation at bandwidth extremes yields comparable achievable rate performance to constant energy waveforms for $\mu = 0.25$ and 0.5, and similar to [4] for $\mu = 1$. The decrease in achievable rate with increasing ROF μ is attributed to the reduced SNR per resource bin while maintaining constant total energy. Finally, Fig. 7 shows that the proposed interpolation method effectively reduces sidelobes in the estimated Channel Impulse Response (CIR). At low ROF ($\mu = 0.25$), prominent sidelobes can affect delay accuracy, but the interpolation method achieves a $2\times$ reduction in sidelobe levels under low resource conditions.

6 Conclusions

Designing ISAC waveforms for 6G networks presents significant challenges, necessitating a balance between high-capacity communication and precise sensing. This work introduces two innovative waveforms: a dual-domain waveform that integrates OFDM with a customized sensing signal in the delay-Doppler domain, and an optimal time-frequency-energy resource allocation strategy combined with a novel interpolation technique using Schatten p-norm matrix completion. These advancements enhance sensing resolution while maintaining communication rate quality of service and address high sidelobes caused by underutilized FT resources. Numerical results show that the proposed designs significantly outperform existing methods, providing substantial benefits for 6G networks by effectively integrating advanced communication and sensing functionalities.

References

1. Baquero Barneto, C., Riihonen, T., Turunen, M., Anttila, L., Fleischer, M., Stadius, K., Ryynä-nen, J., Valkama, M.: Full-duplex OFDM radar with LTE and 5G NR waveforms: challenges, solutions, and measurements. IEEE Trans. Microwave Theory Tech. **67**(10), 4042–4054 (2019). https://doi.org/10.1109/TMTT.2019.2930510
2. ETSI TS 138 104 V15.2.0: 5G; NR; Base Station (BS) radio transmission and reception (3GPP TS 38.104 version 15.2.0 Release 15), July 2018
3. Garcia, M.H.C., Molina-Galan, A., Boban, M., Gozalvez, J., Coll-Perales, B., Şahin, T., Kousaridas, A.: A tutorial on 5G NR V2X communications. IEEE Commun. Surv. Tutor. **23**(3), 1972–2026 (2021). https://doi.org/10.1109/COMST.2021.3057017
4. Keskin, M.F., Koivunen, V., Wymeersch, H.: Limited feedforward waveform design for OFDM dual-functional radar-communications. IEEE Trans. Signal Process. **69**, 2955–2970 (2021). https://doi.org/10.1109/TSP.2021.3076894
5. Liu, F., Zhou, L., Masouros, C., Li, A., Luo, W., Petropulu, A.: Toward dual-functional radar-communication systems: optimal waveform design. IEEE Trans. Signal Process. **66**(16), 4264–4279 (2018)
6. Mura, S., Tagliaferri, D., Mizmizi, M., Spagnolini, U., Petropulu, A.: Optimized waveform design for OFDM-based ISAC systems under limited resource occupancy (2024). arXiv preprint arXiv:2406.19036
7. Mura, S., Tagliaferri, D., Mizmizi, M., Spagnolini, U., Petropulu, A.: Waveform design for OFDM-based ISAC systems under resource occupancy constraint. In: 2024 IEEE Radar Conference (RadarConf24), pp. 1–6 (2024). https://doi.org/10.1109/RadarConf2458775.2024.10548861
8. Raviteja, P., Phan, K.T., Hong, Y., Viterbo, E.: Interference cancellation and iterative detection for orthogonal time frequency space modulation. IEEE Trans. Wirel. Commun. **17**(10), 6501–6515 (2018). https://doi.org/10.1109/TWC.2018.2860011
9. Tagliaferri, D., Mizmizi, M., Mura, S., Linsalata, F., Scazzoli, D., Badini, D., Magarini, M., Spagnolini, U.: Integrated sensing and communication system via dual-domain waveform superposition. IEEE Trans. Wirel. Commun. 1–1 (2023). https://doi.org/10.1109/TWC.2023.3316888
10. Xiong, Y., Liu, F., Cui, Y., Yuan, W., Han, T.X., Caire, G.: On the fundamental tradeoff of integrated sensing and communications under Gaussian channels. IEEE Trans. Inform. Theory **69**(9), 5723–5751 (2023). https://doi.org/10.1109/TIT.2023.3284449